100 CLIMATE ACTIONS FROM CITIES IN ASIA AND THE PACIFIC

JUNE 2021

ASIAN DEVELOPMENT BANK

ADB

Contents

⚡ CLEAN AND RENEWABLE ENERGY

CARBON FINANCE AND PARTNERSHIP

URBAN TRANSPORT AND MOBILITY

LAND USE AND FORESTRY

SMART CITIES

SUSTAINABLE AND LOW-CARBON COMMUNITIES

CLIMATE ACTION PLANS AND INVENTORIES

BUILDING ENERGY EFFICIENCY

SOLID WASTE

CLIMATE RESILIENCE

Foreword

While great strides in economic and social development have been made across developing countries in Asia and the Pacific, the region's increasing share of global greenhouse gas (GHG) emissions and vulnerability to climate risks threaten to derail its development gains. The region is currently home to 20 of the world's 33 megacities with an expected increase to 27 by the year 2030. Rapidly developing cities in particular grapple with inadequate basic services, infrastructure gaps, and environmental degradation.

Amid these challenges lie opportunities for the region's urban centers to balance massive population growth and livability while also building resilience and limiting their carbon footprint. Cities can play a pivotal role by forging partnerships and embarking on integrated approaches to scale up and accelerate climate actions that can contribute to broader sustainability.

The distinctive characteristics of developing Asia's cities reflected through its geography offer urban managers, practitioners, and citizens the opportunity to design and develop adaptable initiatives that can help transform the urban landscape to promote climate resilience and low-carbon development. As economic growth engines and strategy enablers, cities are in a unique position to leverage innovative financing, systems, and technologies to improve competencies and capacities. This publication showcases climate responsive transformations anticipated or already achieved by projects in developing cities in the region with actions ranging from non-traditional nature-based solutions and informed urban planning to strategic carbon finance and smarter technologies.

The Asian Development Bank (ADB) is committed to supporting a regional shift toward a low-carbon and climate-resilient development path. We would like to acknowledge our donor partners who contribute greatly to the Asian Development Fund (ADF) in providing grants to ADB's poorest and most vulnerable countries. As we stay the course toward the transformation of the region's developing cities into safe, inclusive, and sustainable urban centers, we will continue to support local governments as they strengthen climate actions by providing financing and technical assistance that leverages investments from the private sector, public sector, and other development partners.

Ahmed M. Saeed
Vice-President
East Asia, Southeast Asia, and the Pacific
Asian Development Bank

Introduction

Cities in developing countries in Asia and the Pacific are growing at an unprecedented speed. The region's population is expected to increase from 1.84 billion in 2017 to 3 billion by 2050, with a projected urbanization rate of 64%. While cities act as engines of economic growth, rapid urbanization coupled with population growth poses a significant challenge to sustainable development for a region that is the most vulnerable in the world to the impact of climate change.

To cope with this anticipated population growth while adapting to and mitigating climate change, several cities in the region have set ambitious climate targets with the aim of reducing their greenhouse gas (GHG) emissions and strengthening their adaptive capacities.

Drawing from experiences within the region, from multiple sectors including renewable energy, carbon finance, transport, land use, information and communication technology, climate action plans, building energy efficiency, solid waste, sustainable and low-carbon communities, and climate resilience, this publication illustrates how city-level initiatives contribute to reducing GHG emissions and building resilience while delivering economic, environmental, health, and social co-benefits.

These experiences also underscore the key role of cities as initiators, institution builders, and innovators that can translate climate solutions into actions. Importantly, it recognizes stakeholder engagement, collaborations, and partnerships as enablers to accelerate climate actions that are tailored specifically to local conditions.

The Asian Development Bank (ADB) remains steadfast in charting the path toward low-carbon and climate-resilient development. It is our hope that this publication will be useful to urban development stakeholders, enabling developing countries within the region to craft and align with their national strategies innovative and responsive climate solutions that can contribute to building sustainable, inclusive, and livable cities.

James Lynch
Director General
East Asia Department
Asian Development Bank

Bruno Carrasco
Director General concurrently
Chief Compliance Officer
Sustainable Development
and Climate Change Department
Asian Development Bank

Strengthening Kiribati's climate resilience. Kiribati is an island nation in the central Pacific Ocean that faces a range of climate change–related challenges. The country is taking a range of measures aimed at tackling the freshwater shortages, flooding, and prolonged droughts projected in the future (photo by ADB).

Acknowledgments

This publication was conceptualized and prepared under the leadership of ADB's East Asia Department with support from the Clean Energy Fund under the Clean Energy Financing Partnership Facility, Clean Technology Fund, Governance Cooperation Fund, People's Republic of China Poverty Reduction and Regional Cooperation Fund, Regional Cooperation and Integration Fund, and Republic of Korea e-Asia and Knowledge Partnership Fund. Na Won Kim, senior urban development specialist, spearheaded the overall production of the publication, with the guidance and supervision of Sujata Gupta, director of East Asia Sustainable Infrastructure Division and Manoj Sharma, chief of Urban Sector Group.

Dianne April Delfino, Gloria Gerilla-Teknomo, and Ellen May Reynes coordinated the development of this publication. Coordination support from Aigerim Akiltayeva, Chen Chen, Zolzaya Enkhtur, Dorjgotov Otgonbaatar, Patrick Zulla, and the Hunan Innovative Low-Carbon Center was also provided.

Contributions of project information and images from development partners and stakeholders are gratefully acknowledged. They are Baatad Altan-Ulzii, M. Solaiman Bakhshi, Bold-Erdene Bayaraa, Jusup Bekbolotov, Kerrie Burge, Hongjin Cai, Lei Can, Fengbin Chen, Seti Chen, Elene Goksadze, Jamoliddin Holista, Jun Huang, Xiong Jihai, Soóalo Kuresa, Salohiddin Mamadaliev, Ariunjargal Myagmardorj, Lasha Nakashidze, Munkhzolboo Purev, Liu Qian, Sayfullo Qoridov, Eldar Salahov, Surab Secreted, Rahman Shah, Tserendash Sugarragchaa, Yonghe Sun, Tuul Undarmaa, Kien Vu, Simon Wilson, Lubei Yi, Weijun Zhang, and Zhang Zhao.

Greatly appreciated as well were the suggestions and inputs of the following ADB staff and consultants: Rafayil Abbasov, Askar Abeuov, Tuul Badarch, Joy Amor Bailey, Kelly Bird, Stephen Blaik, Vivian Castro-Wooldridge, Christine Chan, Chenglong Chu, Alexandra Conroy, Luca Di Mario, Hoang Nhat Do, Marga Domingo-Morales, Judith Doncillo, David Richard Fay, Maria Vicedo Ferrer, Len George, Bertrand Goalou, Najibullah Habib, Mohammed Azim Hashimi, Shuji Hashizume, Arnaud Heckmann, Won Myong Hong, Cahyadi Indrananto, Rustam Ishenaliev, Satoshi Ishii, Ariel Javellana, Andrew Jeffries, Okju Jeong, Kristina Katich, Sarocha Kessakorn, Tristan Knowles, Yoshiaki Kobayashi, Jaimes Kolantharaj, Cesar Llorens, Xijie Lu, Declan Magee, Ali Rahim Malik, Rosemarie Marquez, Daisuke Mizusawa, Marzia Mongiorgi, Sumika Moriue, Takahiro Murayama, Alexander David Nash, Kiron Nath, Tshewang Norbu, Francis Mark Pascual, Wolfgang Pocheim, Arun Ramamurthy, Markus Roesner, Kristian Rosbach, Tomas Eric Sales, Sharad Saxena, Arman Seissebayev, Pushkar Srivastava, Raquel Rago Tabanao, Momoko Nitta Tada, Samia Tariq, Cindy Cisneros Tiangco, Joris Van Etten, Emma Veve, Johannes Eberhard Vogel, Ruediger Zander, and Baochang Zheng.

We sincerely acknowledge the valuable advice and comments received from the peer reviewers: Frederic Asseline, David Elzinga, Ki-Joon Kim, Virinder Sharma, and J. Michael Trainor.

This publication has immensely benefited from the assessment and design of Sustainia, whose team include Lindsey Chaffin, Lise Kjølbye, Kelly Lynch, Rasmus Schjødt Pedersen, and Jack Robinson.

Abbreviations

BRT

bus rapid transit

CO_2

carbon dioxide

GDP

gross domestic product

GHG

greenhouse gas

ICT

information and communication technology

LED

light emitting diode

MRT

mass rapid transit

NOx

nitrogen oxide

PRC

People's Republic of China

Weights and Measures

GWh

gigawatt-hour

km^2

square kilometer

kWh

kilowatt-hour

MWh

megawatt-hour

tCO_2e

tons of CO_2 equivalent

Urban greening. Green spaces provide a range of benefits to the environment and the urban population (photo by Cao Shengli).

Prologue

→ 100 City Projects for Climate Action in Asia and the Pacific

There is enormous variety across cities in Asia and the Pacific, reflected in the unique challenges and opportunities associated with climate change that each of them face. Despite these differences, it is possible to see a common desire to contribute toward climate action.

This publication clearly demonstrates that no matter the circumstances, cities are in a unique position to take action at the local level to address this global phenomenon. From transitioning to clean cookstoves in Afghanistan, to harnessing volcanic geothermal energy in Indonesia, and from improving access to clean water in Bangladesh, to vast flood resilience plans in the People's Republic of China (PRC), the region is increasingly turning away from dirty and limited fossil fuels to cleaner and renewable energy sources.

Cities in Asia and the Pacific are also some of the most vulnerable to climate change, with many already feeling the impact of rising temperatures. Low-lying Pacific islands face disruptive sea level rise predictions, and rapidly growing megacities are challenged with providing urban services under stressed climatic conditions. As disasters become more frequent and severe under the warming climate, cities in the region have no option but to adapt and become more resilient. These cities are looking past the challenges to see the opportunities associated with cleaner air and water, lower congestion, and improved waste treatment. Win-wins are a commonality in these climate action cases.

This publication demonstrates some of the efforts of the Asian Development Bank (ADB), other development partners, governments, and the private sector to support cities to address climate change and showcase their innovation in low-carbon city development. ADB hopes that by sharing these examples, other cities will be inspired to drive further innovation and transform their cities to protect against climate change.

CITY PROJECTS IN THIS PUBLICATION ARE DIVIDED INTO 10 SECTORS

Clean and Renewable Energy

Carbon Finance and Partnership

Urban Transport and Mobility

Land Use and Forestry

Smart Cities

Sustainable and Low-Carbon Communities

Climate Action Plans and Inventories

Building Energy Efficiency

Solid Waste

Climate Resilience

NAVOI

Harnessing Sunshine in Uzbekistan
for the First Time
p. 6

KYZYLORDA AND SHU

Energizing Kazakhstan's South
with Solar
p. 26

TASH-KUMYR

Updating the Kyrgyz Republic's Aging
Hydropower
p. 21

BAKU

Ramping Up Renewables
atop a Fortune of Fossil Fuels
p. 5

KANDAHAR

Afghanistan's First Utility-Scale
Solar Plant
p. 19

DELINGHA

Concentrated Mitigation Efforts
p. 14

DHAKA

Utility-Scale Solar Comes to Dhaka
p. 20

VISAKHAPATNAM

Water and Power from
Visakhapatnam's Reservoir
p. 9

PADANG AND PAGAR ALAM

Tapping Into Indonesia's Natural
Heat Potential
p. 10

Clean and Renewable Energy

→ All across Asia and the Pacific, countries are exploring the previously untapped potential of renewable energy sources and beginning to transition away from fossil fuels. By harnessing the power of increasingly cost competitive wind, solar, hydro, geothermal, and biomass energy, cities are making progress toward ambitious renewable energy targets and, in some cases, reducing dependence on costly energy imports.

Fueling Cars with Corn in the PRC

The northern city of Tieling is scaling up corn-based ethanol production to provide Chinese consumers with biofuel to reduce transport emissions in the People's Republic of China (PRC).

A new production facility in the city of Tieling, 8 hours northwest of Beijing, is helping to meet the PRC's targets for biofuel consumption. Via a series of complex reactions starting with corn, the facility produces 300,000 tons of ethanol fuel, 276,300 tons of high-protein livestock feed, and 20,000 tons of corn oil every year.

The ethanol produced can be mixed with regular gasoline for a cleaner-burning transportation fuel, which the Government of the PRC has prioritized as a target to improve urban air quality.

Since 2020, the PRC has required that gasoline supplies nationwide be blended with 10% ethanol (also known as E10 fuel, common elsewhere in the European Union and the United States), which entails the production of around 15 million tons of the biofuel annually.

↑300K

TONS OF ANNUAL BIO ETHANOL FUEL PRODUCTION

Inhabitants*
2.89 million

Gross domestic product (GDP) per capita*
$3,530

Geographic area*
12,966 square kilometers (km²)

* Unless otherwise indicated, all information represents city-level data.

THE CHALLENGE

Corn-based ethanol production is helping to address the province's backlog of corn, the hoarding of aged grain, and the farmers' logistics and storage expenditure.

CO-BENEFITS

Economic

The increased demand for corn will improve the outlook for local farmers around Tieling.

Health

Utilizing E10 fuel can reduce polluting particulate emissions from tailpipes by around 20%, helping to improve air quality.

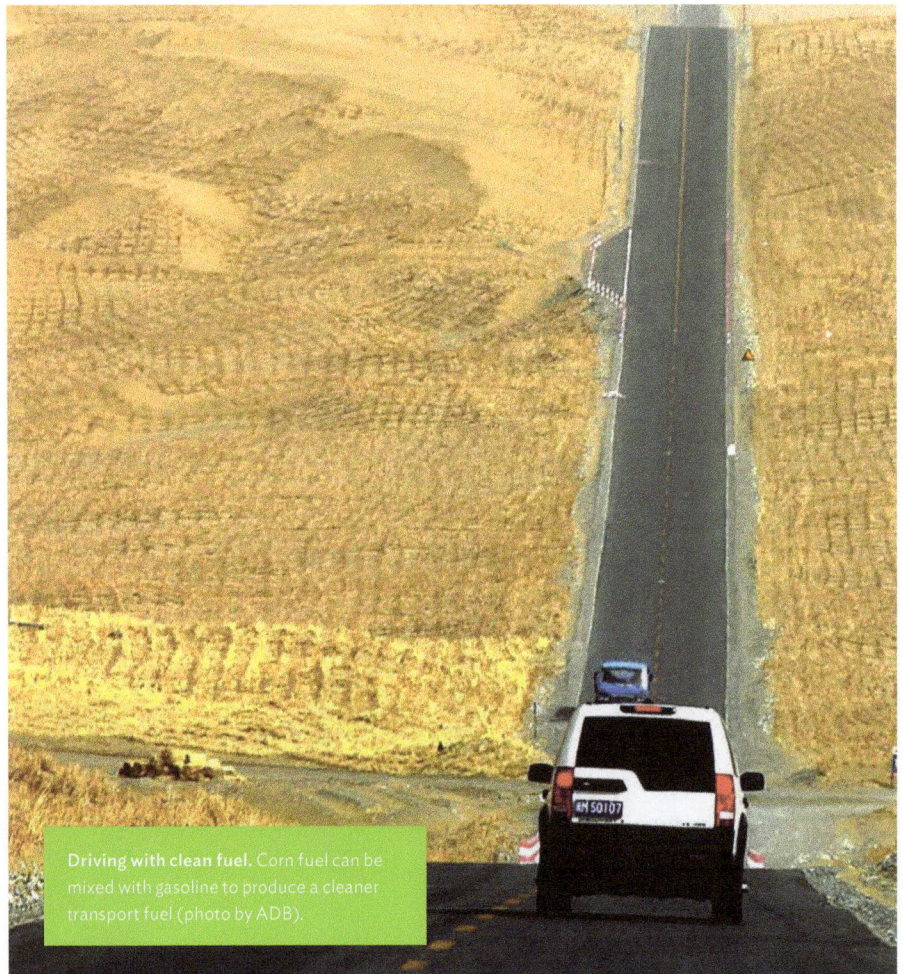

Driving with clean fuel. Corn fuel can be mixed with gasoline to produce a cleaner transport fuel (photo by ADB).

Ramping Up Renewables atop a Fortune of Fossil Fuels

↓**3.9M**

TONS OF CO_2 EMISSIONS
REDUCED BY 2030

Inhabitants
2.29 million

GDP per capita
$4,689

Geographic area
2,140 km²

THE CHALLENGE

As the capital and largest city of Azerbaijan, Baku strives to contribute significantly toward the country's achievement of its ambitious renewable energy targets.

CO-BENEFITS

Economic

By contributing to energy security, the project helps to protect against energy unreliability, preventing economic losses.

Social

Developing solar and other renewables will create green jobs for the people of Azerbaijan.

Despite being a major crude oil and natural gas producer, Azerbaijan is taking the first steps to transition toward a more sustainable power sector with ambitious renewable energy targets.

Azerbaijan aims to triple its current capacity of wind, solar, biomass, and hydro electricity generation to give renewables a 30% share of the country's total installed power capacity by 2030. Energy diversification efforts have been underway since 2013, which have scaled renewables to account for 7.3% of the electricity production, thus far.

This ambition is especially notable since Azerbaijan is among the top 25 oil and gas-producing countries. The planned solar, wind, biomass, and hydro developments are expected to reduce, by 2030, a total of 3.9 million tons of carbon dioxide equivalent (tCO_2e) emissions from the energy sector, which is by far the country's largest source of emissions.

In addition to wind and land-based solar power, floating solar photovoltaic (FPV) power has been identified as an area of potential, and Lake Boyukshor is one of the first pilot sites in Central Asia for this technology. The previously polluted saline lake is the largest of nine lakes in Azerbaijan's Absheron peninsula and will soon host a 100-kilowatt FPV system. The pilot project is implemented through technical assistance from ADB, which will also explore the feasibility of a scale-up plant, provide technical capacity building, and develop detailed business models to encourage private sector participation. Located in the Azeri capital of Baku, the technical assistance will pilot test high-level technology, showcasing the potential of innovative FPV development to local universities, research institutions, private sector companies, and the general public.

Azerbaijan's clean energy mix.
Azerbaijan is aiming to triple its renewable energy generation capacity by 2030 (photo by ADB).

Harnessing Sunshine in Uzbekistan for the First Time

↓140K

TONS OF CO$_2$ EMISSIONS REDUCED ANNUALLY

Inhabitants
467,600

GDP per capita
$1,251

Geographic area
35 km^2

THE CHALLENGE

Largely covered by the Kyzylkum desert, Navoi's potential as a viable site for large-scale solar park development has not been tapped fully.

CO-BENEFITS

Economic

By contributing to energy security, the project will help to protect against energy unreliability, preventing economic losses.

Social

Developing solar and other renewables will create green jobs for the people of Uzbekistan.

Uzbekistan is rich in sunshine hours, yet 90% of its electricity is sourced from fossil fuels. The country's first solar park will initiate a transition to renewables for electricity needs and push this number down.

A new PV solar park will be established in Navoi, a region in the north of Uzbekistan mostly covered by the Kyzylkum desert. The region will host what is hoped to be the first of multiple large-scale solar parks, with a starting capacity of 100 megawatts (MW) and an estimated production output of 255 million kilowatt-hours (kWh) per year, enough to power 150,000 homes. It will also reduce an estimated 140,000 tCO$_2$e emissions every year through the displacement of fossil fuel generation.

The power plant will start supplying the grid in the first quarter of 2022 and will reshuffle Uzbekistan's energy portfolio and grow the share of renewables beyond the current 10% that is held by hydropower. With a goal to bring renewables' share to 25% by 2030 on the national agenda, 9% is planned to be held by solar power, with the remaining 11% to be held by hydropower and 5% by wind power.

The project received a $13 million loan from ADB and an $8 million loan from the ADB-administered Canadian Climate Fund.

Uzbekistan's clean energy target.
Uzbekistan is aiming to have 25% renewable energy by 2030 (photo by ADB).

Hydropower Makes a Splash in Solomon Islands

↑68%

OF HONIARA'S POWER DEMAND MET THROUGH THE PROJECT

Inhabitants
84,520

GDP per capita
$2,197

Geographic area
22 km²

THE CHALLENGE

Solomon Islands' capital city Honiara is home to 84,000 people, more than 10% of the country's population, but until now has been powered almost entirely by diesel fuel.

CO₂ BENEFITS

Economic

This project, combined with a push for more solar, will lower power prices for homes and businesses across the country, where the price of electricity is among the highest in the world.

Social

The project will contribute to poverty reduction, improve the reliability of electricity for existing customers, and provide employment during project construction and implementation.

Health

Reduced emissions from diesel combustion is expected to lower the risk of cardiovascular and cardiopulmonary diseases through improvements in air quality.

The residents of Honiara, the capital city of Solomon Islands, will benefit from a new 15-megawatt (MW) hydropower plant on the Tina River that will help replace diesel and boost renewable electricity to 85% of the mix for the city.

The project's new 15 MW hydropower plant on the Tina River is the largest ever public–private partnership in the country, with the state utility entering into a 34-year power purchase agreement for the electricity generated. The project is due for completion by the end of 2024.

The Tina River project will provide an estimated 68% of the power demands from Honiara and will also curb Solomon Islands' reliance on imported diesel by almost 70%. With the commissioning of the plant and proposed solar investments, the extent of renewable energy integration is expected to grow from 1% in 2016 to 85% in 2022 (68% from the hydropower plant and 17% from solar). The Tina River project is expected to reduce CO_2 emissions by around 50,000 tCO_2e per year.

The project is funded through an ADB-administered grant ($12 million) and loan ($18 million); cofinancing through the Abu Dhabi Development Fund ($15 million), Government of Australia ($11.70 million), Green Climate Fund ($86 million), Export-Import Bank of Korea ($31.60 million), World Bank ($31.20 million); and government cofinancing ($17.07 million).

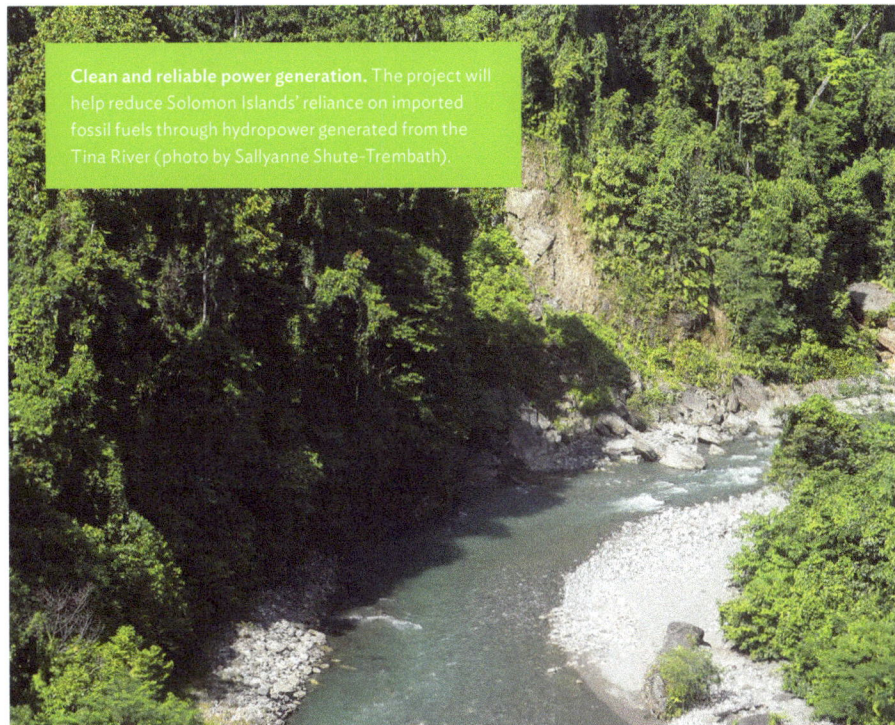

Clean and reliable power generation. The project will help reduce Solomon Islands' reliance on imported fossil fuels through hydropower generated from the Tina River (photo by Sallyanne Shute-Trembath).

↑**22.3**

GWH OF ELECTRICITY
PRODUCED ANNUALLY

Inhabitants
1.54 million

GDP per capita
$5,653

Geographic area
4,704 km²

THE CHALLENGE

With Ulaanbaatar's large population and heavy reliance on fossil fuels, the Government of Mongolia needs to shift to clean energy sources and increase its share of renewable energy to a targeted 30% by 2030.

CO-BENEFITS

⟍ Economic

Installing new PV capacity leads to generation of local employment.

♡ Health

Reducing Mongolia's reliance on coal for energy will improve air quality especially in densely populated cities like Ulaanbaatar.

ULAANBAATAR, MONGOLIA

Making the Most of the Sun in Mongolia

Despite freezing winters, Mongolia's capital has bountiful solar radiation, which the 15 MW solar plant just outside Ulaanbaatar is now harvesting.

On an open grassland steppe 40 kilometers (km) from Mongolia's capital city lies one of Mongolia's largest solar power plants—a 15 MW array with over 15,000 PV panels. It provides an estimated 22.3 gigawatt-hours (GWh) of electricity annually that is fed into the national grid, while reducing the country's carbon emissions by 26,400 tons annually.

Ulaanbaatar is the coldest capital city in the world. Despite this, there is still much potential for solar energy. The city has an annual average of 2,800 hours of sunshine, which is more than Madrid's average.[1] Ulaanbaatar also struggles with some of the world's worst air pollution in the winter, although this is mostly due to burning coal for heat and from the transportation sector.

This is one of several projects aiming to develop Mongolia's renewable energy portfolio and help increase the share of renewable energy in the total installed power generation capacity to 20% by 2023 and 30% by 2030.

The $26.7 million project is financed through loans from ADB and the ADB-administered Leading Asia's Private Infrastructure Fund.

---

[1] Weather and Climate weather-and-climate.com.

Green jobs in Ulaanbaatar. Local green jobs are created with the push for renewables in Mongolia (photo by ADB).

Water and Power from Visakhapatnam's Reservoir

A new floating solar power project on the Meghadrigedda reservoir is underway in the outskirts of Visakhapatnam in India.

The Greater Visakhapatnam Municipal Corporation is installing an innovative floating solar power plant on the Meghadrigedda reservoir, which is also an important source of freshwater for the city. Upon completion in 2022, the 3 MW plant will provide renewable energy to the grid, averting 1,500 tCO_2e annually. The project also builds on a previous nearby floating solar project on Mudasarlova reservoir, where both the panels as well as the floating devices have been improved. The floating plant is not expected to affect the ecosystem of the reservoir or the city's drinking water, and will take up only 1% of reservoir space.

The floating solar project is part of the larger Visakhapatnam–Chennai Industrial Corridor Development Program and has received a $5 million grant from the ADB-managed Urban Climate Change Resilience Trust Fund. This floating solar project is also running alongside a number of climate adaptation projects focusing on watershed management and rejuvenation of the Mudasarlova reservoir.

↓1.5K

TONS OF CO_2 EMISSIONS REDUCED ANNUALLY

Inhabitants
2.08 million

GDP per capita
$2,100

Geographic area
567 km²

THE CHALLENGE

Visakhapatnam is one of India's fastest-growing urban centers and vulnerable to climate change-induced natural hazards such as cyclones and storm surges.

CO-BENEFITS

Economic

The project will complement the ongoing efforts of the Government of Andhra Pradesh to enhance manufacturing sector growth and provide clean energy jobs.

Environmental

This floating solar project is complemented by other projects focusing on watershed management and rejuvenation of the Mudasarlova reservoir catchment.

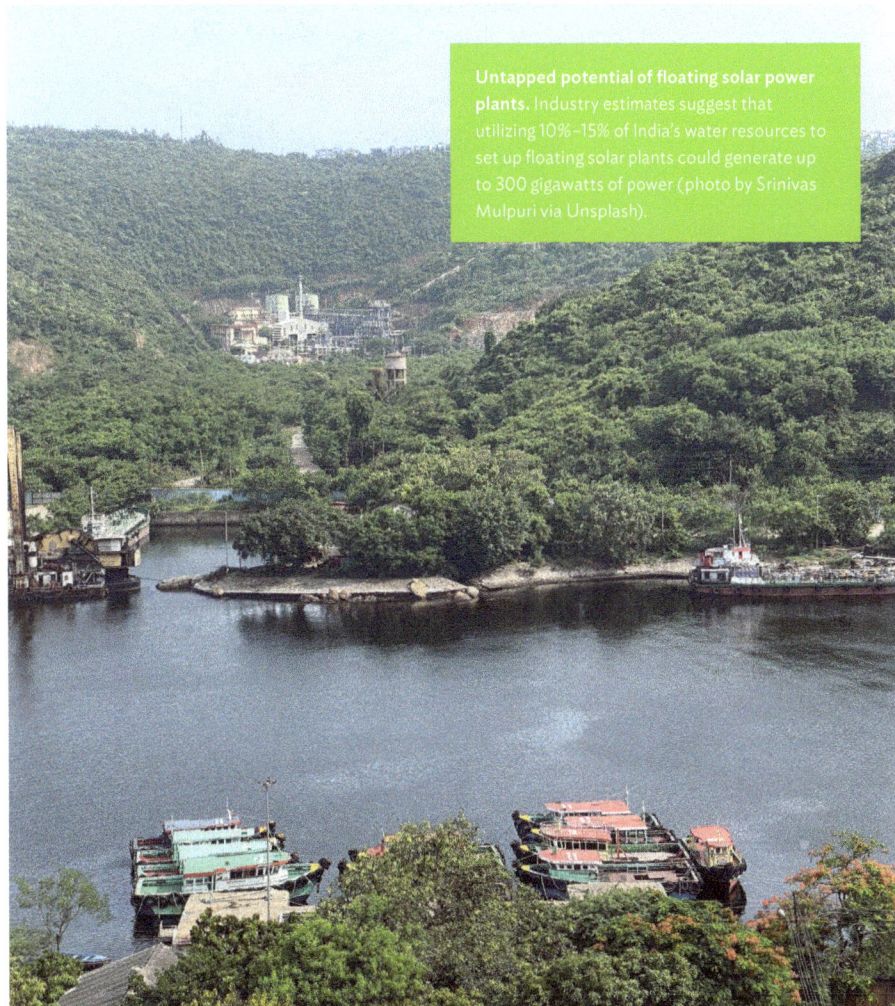

Untapped potential of floating solar power plants. Industry estimates suggest that utilizing 10%–15% of India's water resources to set up floating solar plants could generate up to 300 gigawatts of power (photo by Srinivas Mulpuri via Unsplash).

Tapping Into Indonesia's Natural Heat Potential

Two new power plants with a combined 170 MW of power in Sumatra will be mining energy from Indonesia's rich geothermal resources, to help meet the country's long-term renewable energy goals.

Among the volcanic mountains and tropical rainforests of the Indonesian island of Sumatra lie two of the country's newest geothermal plants. The 80 MW plant close to Padang and the 90 MW plant near Pagar Alam will reduce a combined estimated 870,000 tCO_2e emissions every year.

On the edge of the Pacific Ocean's Ring of Fire, Indonesia is one of the world's most tectonically active regions. The country has an estimated 40% of the world's total geothermal energy capacity, but in 2016 was only harnessing 5% of the total. In an effort to decarbonize the economy and meet the goal of reducing emissions by 29% by 2030 compared to business as usual, the country is ramping up efforts in geothermal energy production.

ADB provided loans totaling over $240 million and administered cofinancing from other sources for the two projects, which resulted in a combined cost of over $1.3 billion.

↓870K

TONS OF CO_2 EMISSIONS REDUCED ANNUALLY

Inhabitants
Padang: 950,870
Pagar Alam: 146,973

GDP per capita
Padang: $4,650
Pagar Alam: $1,420

Geographic area
Padang: 695 km²
Pagar Alam: 634 km²

THE CHALLENGE

Geothermal power generation is the dominant alternative to displace fossil fuel-generated power in Sumatra's grid, but exploration costs and the risks of proving and managing the geothermal steam resources are high.

CO-BENEFITS

Economic

Geothermal power is constant, unlike variable renewable energies like solar or wind power, so it can act as a baseline power supply for Indonesia.

Health

Geothermal power has the added benefits that it does not lead to the same levels of harmful air pollution as fossil fuel-based power generation.

Social

The project is expected to generate new jobs and provide additional income sources for the community.

Geothermal power generation in Sumatra. Padang's new 80 MW geothermal plant went online at the end of 2019 and construction of the Pagar Alam plant will be finished in 2021 (photo by ADB).

Cambodia's First Utility-Scale Solar Plant

Cambodia's first utility-scale solar plant in the southeast of the country is the first to tap into a powerful natural resource in Cambodia—sunshine.

The 10 MW solar power plant is located in Bavet, a special economic zone on the border of Viet Nam, about 150 km from the capital, Phnom Penh. Following an international tender, a consortium led by Singaporean company Sunseap built the plant and entered into a 20-year solar power purchase agreement with Electricité Du Cambodge, a state-owned utility. The solar plant has been providing around a quarter of the demand for nearby city Bavet since 2017 and is estimated to reduce around 5,500 tCO_2e emissions annually.

The local area used to suffer from power shortages. Since the operation of the solar plant, investments in the area have increased, promoting further development and providing employment opportunities for local communities.

ADB's Private Sector Operations Department provided a debt financing package of $9.2 million for the project. The package included cofinancing from a private sector financial institution and a concessional loan from the ADB-managed Canadian Climate Fund for the Private Sector in Asia (CFPS).

↓5.5K

TONS OF CO_2 EMISSIONS REDUCED ANNUALLY

Inhabitants
42,456

GDP per capita
$1,621

Geographic area
207 km²

THE CHALLENGE

Bavet, and Cambodia more broadly, has historically relied on fossil fuels and hydropower for energy, despite great untapped potential for solar power.

CO-BENEFITS

Economic

Since the solar plant started generating power, investments in the local economic zone have increased, promoting development in the area.

Social

The solar project has improved the reliability of Cambodia's power supply and has led to the creation of green jobs.

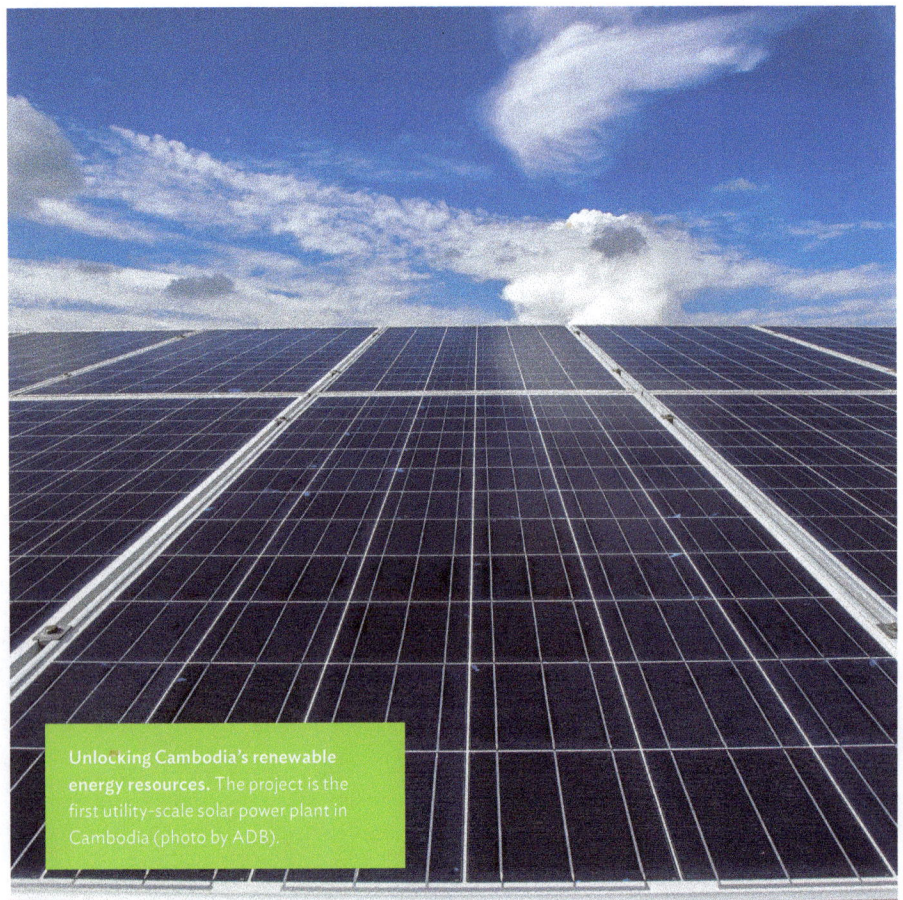

Unlocking Cambodia's renewable energy resources. The project is the first utility-scale solar power plant in Cambodia (photo by ADB).

Micro-grids for the Federated States of Micronesia

↑30%
OF DEMAND WILL BE MET THROUGH RENEWABLE ENERGY

Inhabitants
Kosrae: 7,686
Yap: 16,436

GDP per capita
Kosrae: $2,309
Yap: $3,388

Geographic area
Kosrae: 111 km²
Yap: 118 km²

THE CHALLENGE

Imported diesel is the main fuel on the FSM, accounting for about 80% of power generation on Yap and 95% on Kosrae.

CO-BENEFITS

Economic

The increased renewable energy capacity will help to reduce the dependence on imported and expensive fossil fuels.

Social

The provision of social services, such as health and education, is expected to improve in Yap and Kosrae through a more consistent power supply.

The Federated States of Micronesia (FSM) is investing in solar micro-grid and battery energy storage systems as well as capacity building to improve the utility company's commercial viability and financial sustainability and reduce emissions.

On the island of Kosrae, 1.15 MW of grid-connected solar PV capacity is being installed as well as solar–diesel hybrid mini-grid and rooftop solar systems for homes. On Yap, another pristine island on the other side of the FSM's territory, the project will install a 1.95 MW ground-mounted PV solar system together with a battery storage system. The program will also support capacity building with the utility company on the largest island Pohnpei.

The expansion and further integration of renewables on the islands builds on the experience of the recently completed Yap Renewable Energy Development Project to further decrease Yap's reliance on diesel for power generation and reduce GHG emissions from the islands. In the first year of the project, it is expected that renewable energy generation will make up about 30% of demand for Kosrae and Yap, in line with national targets.

The $15.51 million project was made possible with a $15 million grant from ADB and $510,000 million government cofinancing.

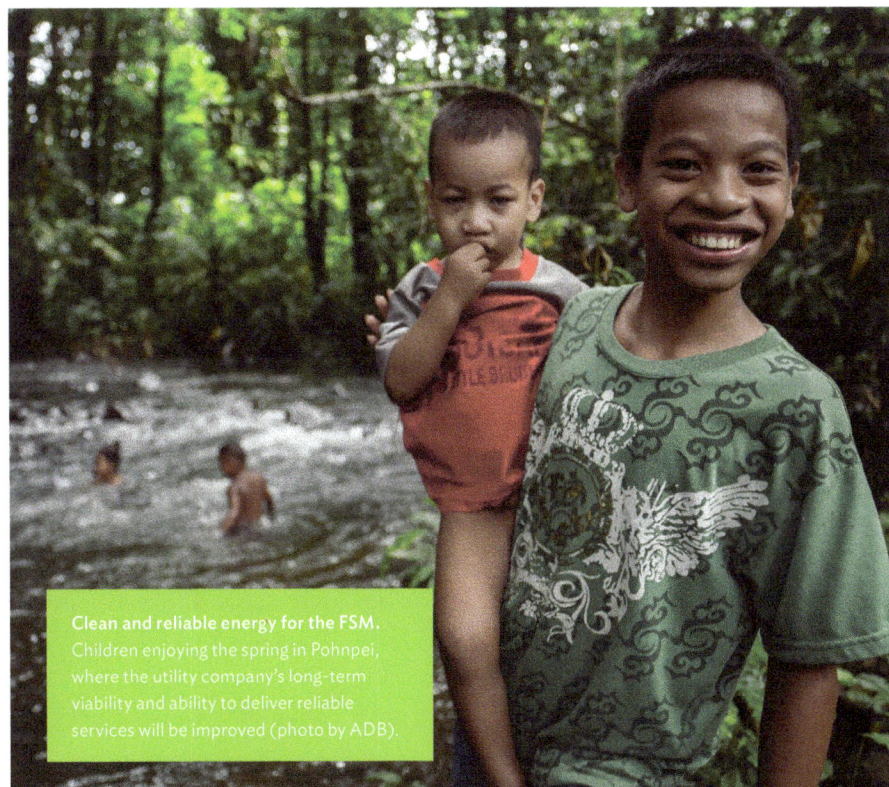

Clean and reliable energy for the FSM. Children enjoying the spring in Pohnpei, where the utility company's long-term viability and ability to deliver reliable services will be improved (photo by ADB).

↓195K

TONS OF CO$_2$ EMISSIONS
REDUCED ANNUALLY

Inhabitants
78,184

GDP per capita
$6,086

Geographic area
27,700 km^2

THE CHALLENGE

Delingha has an abundance
of solar resources that can
propel the growth of a solar-
based industry, but the city
is constrained with limited
financial resources and hands-
on technology experience.

CO-BENEFITS

Economic

Employment opportunities will
help to diversify and boost the
local economy.

Social

This project is contributing
to diversifying the local
economic base as well as
creating 334 unskilled jobs
annually during construction
and 30 permanent jobs during
operation and maintenance.

Concentrated Mitigation Efforts

Delingha's new Concentrated Solar Power (CSP) project is
providing a boost of 50 MW of clean energy to the power grid and
reducing CO$_2$ equivalent emissions by an estimated 195,000 tons
every year.

The 50 MW CSP plant in the city of Delingha, Qinghai Province, is the first of
its kind in the PRC and came into operation in May 2020. The plant works using
parabolically shaped panels that concentrate sunlight to heat water to steam,
which in turn drives a turbine to create power. The thermal element of the design
allows the plant to be combined with energy storage facilities, thus providing
greater flexibility compared with traditional wind or solar energy. The plant will
reduce an estimated 195,000 tCO$_2$e emissions per year.

As well as constructing the first-of-its-kind utility-scale CSP plant in Qinghai
Province, the project also includes capacity development and training in CSP
design, construction, and operation and management.

The project was made possible through an ADB loan of $150 million and
$96.68 million in counterpart financing from the China General Nuclear Power
Corporation (CGN).

Concentrated solar power technology. The
parabolically shaped panels concentrate light
to heat water to produce steam, which in turn
drives a turbine to create electricity (photo by
CGN Delingha Solar Energy, Co. Ltd.).

Small Island Nation, Big Clean Energy Plans

↓13K

TONS OF CO₂ EMISSIONS
REDUCED ANNUALLY

TONGA

Inhabitants
99,600

GDP per capita
$4,364

Geographic area
747 km²

THE CHALLENGE

While 97% of households in urban areas in Tonga have access to electricity, about 90% of this power generation is from imported diesel. Transitioning to cleaner energy sources will help the government meet their 70% renewable energy target by 2030.

CO-BENEFITS

Economic

The project will help to reduce the cost of levelized electricity from $0.40 to $0.30 in Nauru by reducing the reliance on expensive, imported fuel.

Social

The renewable energy project will increase clean energy access of marginalized populations by about 3% in the outer islands, where access is limited.

Tonga's renewable energy project is combining investment in grid-connected renewable energy generation, battery energy storage systems, hybrid systems, and capacity building.

The $53.2 million project will install 650 kilowatts (kW) of grid-connected solar PV capacity with 1.4 megawatt-hours (MWh) battery storage on the islands of 'Eua and Vava'u, and mini-grids totaling 501 kW PV with 4.3 MWh storage in the five outer islands of O'ua, Tungua, Kotu, Mo'unga'one, and Niuafo'ou.

To absorb additional 22 MW of PV and wind power systems to be funded by independent power producers, the project will also install the country's first large-scale battery systems in the capital Nuku'alofa. The batteries, with a total capacity 19.9 MWh, will store excess renewable energy to supply demand when the sun is not shining. This smart use of battery systems will enable Tonga to increase renewable energy penetration close to 50% nationwide without negatively affecting the island grids.

The investments will avoid over 13,000 tCO₂e emissions annually. The final project component focuses on capacity building, which will help the country to better plan for the green transition to reduce reliance on imported fossil fuels and encourage private investment in renewable energy projects.

The project was funded through an ADB grant of $12.2 million, a Government of Australia grant of $2.5 million, a Green Climate Fund grant of $29.9 million, and government counterpart financing of $8.6 million.

Increasing clean energy access. Tonga has a total of 1,151 kW of new grid-connected photovoltaic capacity and over 20 MWh of battery energy storage on the main island of Tongatapu and seven outer islands (photo by Andrew Richard Kautoke).

Harnessing as Much as Land Enables in Kaihua County

After 3 years of implementing several projects within solar, wind, and biomass energy, Kaihua County in Quzhou is striving to become a low-carbon leader.

Located in the westernmost portion of Zhejiang province, Quzhou was recently occupied with new energy development projects. From 2016 to 2018, several projects were raising the local share of renewable energy using solar PV and thermal energy, wind energy, and biomass.

In terms of solar energy, PV power has been a strong focus, with a total of 136 MW of additional capacity installed to date, with rooftop, agricultural, and dispersed installations occurring in the region. Solar thermal has also been targeted, with a total estimated area of 16,000 square meters of panel coverage—roughly two times as big as a Manhattan city block.

Wind and biomass energy are also being harnessed by Quzhou at the moment, with the 60 MW of newly installed wind power and biomass utilization saving an estimated 30,000 tons of coal in total.

↑30K
TONS OF COAL SAVED

Inhabitants
255,500

GDP per capita
$8,701

Geographic area
2,237 km²

THE CHALLENGE

Energy consumption in Quzhou relies heavily on coal.

CO-BENEFITS

♡ Health

Displacing fossil fuel energy generation with renewable energy will contribute to an improved local air quality and corresponding health of citizens.

Environmental

In addition to the renewable energy developments, the project is also establishing a reforestation area of 62,400 acres in size.

Renewable energy as a priority in Quzhou. Photovoltaic solar power has been a big focus for Quzhou, where they have installed 136 MW to date (photo by Rui Jia).

Heating Homes without the Smog

↑10K

APPLICATIONS FOR INSTALLATIONS

Inhabitants
1.54 million

GDP per capita
$5,653

Geographic area
4,704 km²

THE CHALLENGE

Coal use is still widespread in Ulaanbaatar, particularly for residents living in off-grid traditional nomadic dwellings called *ger*. This is a leading cause of air pollution in the Mongolian capital.

CO-BENEFITS

⟋ Economic

To help with costly installation fees, the government has provided green loans to 300 households, and GASCOM is adding to this effort, to help more households make the transition from coal to gas.

♡ Health

Gas is a cleaner-burning fuel alternative to coal that can help communities transition to renewables and improve air quality, reducing the associated health risks such as respiratory diseases.

Mongolia's capital is exploring alternatives to coal to heat homes throughout the harsh winters, reducing the smog that blankets the city during winter months.

The city of Ulaanbaatar is taking a range of measures to reduce the reliance on household coal consumption that is currently the biggest cause of air pollution in the city.

To complement this approach, the Mongolian gas company, Gas Service Corporation of Mongolia (GASCOM), is offering affordable loans to low-income residents to replace coal burners with gas. GASCOM is offering residents the option to change from coal to gas-fired stoves for heating and water. The company is also retrofitting larger buildings including schools and hospitals.

Although still using fossil fuels and not a sustainable long-term solution, gas-fired stoves can help the capital city reduce its reliance on coal and lead to GHG reductions. Burning coal in homes is one of the leading causes of air pollution, which reaches dangerous levels in the winters when temperatures can reach −40°C and demand for energy is highest. The company reports to have already received more than 10,000 applications for the gas-fired equipment.

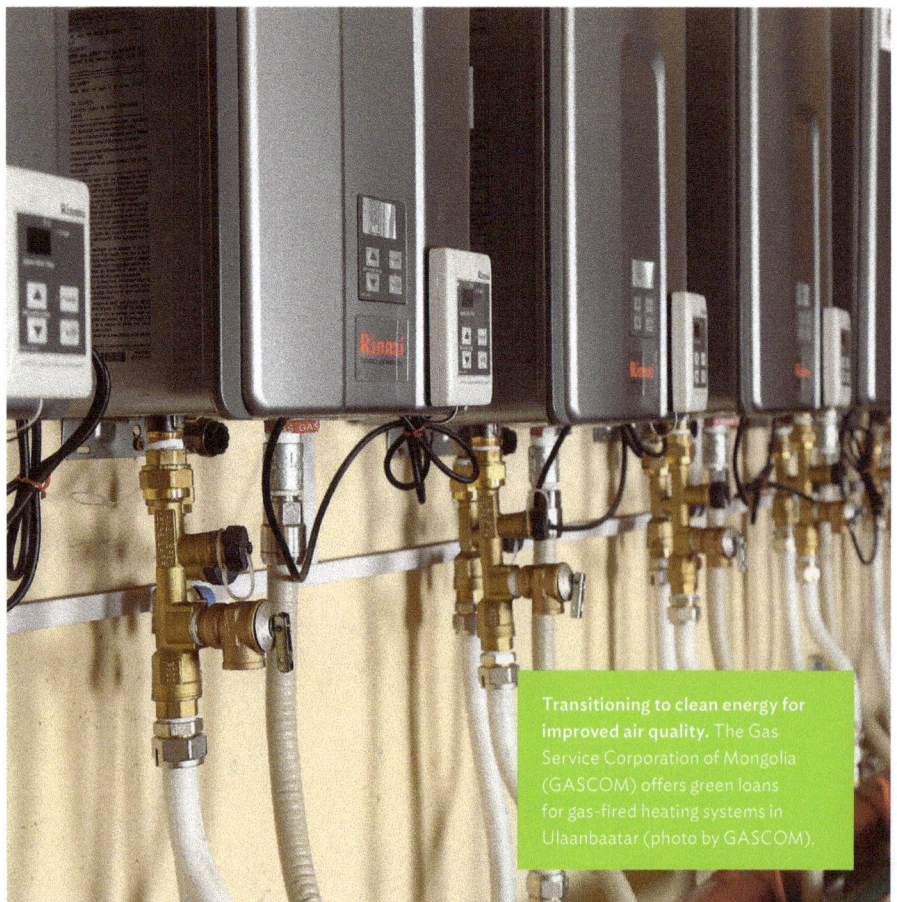

Transitioning to clean energy for improved air quality. The Gas Service Corporation of Mongolia (GASCOM) offers green loans for gas-fired heating systems in Ulaanbaatar (photo by GASCOM).

Afghanistan's First Utility-Scale Solar Plant

The first utility-scale solar plant has been built just outside Afghanistan's second-largest city. The 15 MW plant is the first of many as the country moves toward the 40% renewable target it has set for itself by 2032.

Afghanistan's first and largest solar power plant to date is located 25 km southwest from Kandahar, a 15 MW array with over 55,000 PV panels. It provides an estimated 27.5 GWh of electricity annually fed into the national grid, which is estimated to reduce carbon emissions by over 8,500 tCO_2e annually.

Afghanistan is largely reliant on imported energy, with 80% of the power and 97% of fuel coming from outside its borders. In terms of the energy mix, over half comes from hydropower, and the rest is split between diesel or heavy fuel-powered plants. Before this project, just 3 MW (1%) was coming from solar power, despite the vast capacity the country holds. Renewable energy potential in the country is estimated to exceed 300,000 MW, around six times the level of currently installed generation capacity.

ADB signed a loan with a special purpose vehicle and subsidiaries owned by the 77 Construction, Contracting, and Trading Group (77 Group), and also administered a $3.85 million loan from the Canadian Climate Fund for Private Sector in Asia II (CFPS II) for the project.

↓8.5K

TONS OF CO_2 EMISSIONS REDUCED ANNUALLY

Inhabitants
614,118

GDP per capita
$547

Geographic area
273 km^2

THE CHALLENGE

Kandahar is a rapidly industrializing area and the second-largest city in Afghanistan. It is essential that electricity infrastructure for renewable energy is developed so that it can contribute to reducing Afghanistan's heavy reliance on imported energy.

CO-BENEFITS

Economic

The solar plant is on the periphery of a rapidly industrializing area close to the airport, where the manufacturing sector will benefit from additional electricity supply.

Social

Construction and operation of the Kandahar solar plant has generated employment opportunities for the local community.

A 15 MW solar power plant in Afghanistan. This project is the first of several that will contribute toward the government's target of boosting renewables to a 40% share of the national energy mix by 2032 (photo by 77 Afghanistan).

Utility-Scale Solar Comes to Dhaka

↓30K

TONS CO$_2$ EMISSIONS
REDUCED ANNUALLY

Inhabitants
17 million

GDP per capita
$7,712

Geographic area
2,161 km²

THE CHALLENGE

Dhaka is adversely affected
by increasing gas supply
shortages. While there are
initiatives to promote solar
power, developing grid-
connected solar projects
means overcoming hurdles of
land acquisition and rights of
way for transmission access.

CO-BENEFITS

Economic

Electricity supplied from the
solar power will increase the
availability and reliability of
power, thereby increasing
industrial and agricultural
productivity.

Social

Green employment
opportunities have been
generated during the building
and operation phases for the
local community.

Bangladesh has begun to transition away from its fossil fuel-
reliant energy system of the past, with the installation of a
35 MW solar plant southwest of the country's capital city Dhaka.

Bangladesh's largest solar power plant is located 85 km southwest of Dhaka,
the country's densely populated capital city. It has been completed in 2020.
With 35 MW of power, the project is estimated to provide enough electricity for
around 500,000 households and reduce carbon emissions by over 30,000 tCO$_2$e.

To learn from this project as Bangladesh scales up its renewable energy capacity,
the benefits to local communities will be reported on an ongoing basis. Lessons
from this project, especially in terms of public–private partnerships may be applied
elsewhere given the country's renewable energy expansion.

The financing comprises a loan from ADB and a loan from the ADB-administered
Canadian Climate Fund for the Private Sector in Asia II (CFPS II).

Renewable energy expansion in Bangladesh. The 35 MW solar power plant contributes to the country's target for renewable energy generation capacity of 10% of total power generation by 2021 (photo by Spectra Solar Park Limited).

Updating the Kyrgyz Republic's Aging Hydropower

The Uch-Kurgan hydropower plant is receiving a face-lift after 60 years of operation to boost the power output and improve resilience to climate change.

From 270 km southwest of the capital city Bishkek, the Uch-Kurgan hydropower plant has been providing clean energy to residents of the mountainous Kyrgyz Republic for over 60 years. However, with aging electrical and mechanical equipment, it is long overdue for a modernization.

A $160 million redevelopment project will do just that by updating the generating units and reinforcing the infrastructure to boost output capacity from 180 MW to 216 MW. This is expected to increase energy output by around 15% and lead to emissions reduction of around 11,000 tCO_2e annually. The project will be funded through an ADB loan, and cofinancing from the Eurasian Development Bank and the government.

The program is also taking the chance to enhance resilience to climate change. Repairing the hydraulic steel structures and strengthening structures in the reservoirs can help to prevent flooding. Sediment dredging will also enhance water supply and irrigation.

↓11K

TONS CO_2 EMISSIONS REDUCED ANNUALLY

TASH-KUMYR

Inhabitants
34,756

GDP per capita
$698*

Geographic area
47 km²

*regional data

THE CHALLENGE

To increase the availability of clean hydropower for domestic use and potential export to neighboring countries while contributing to multiyear water supply, immediate major rehabilitation and replacement must be undertaken for the Uch-Kurgan hydropower plant.

CO-BENEFITS

Economic

The project will help strengthen the Kyrgyz Republic's energy self-sufficiency and increase its potential for renewable energy exports to neighboring countries in Central Asia.

Social

The redevelopment of the hydropower plant will increase the safety of local residents who often use the reservoir area for swimming.

Increasing hydropower contribution in the Kyrgyz Republic's energy supply. In 2018, about 80% of the Kyrgyz Republic's 3,920 MW of electricity generation capacity was from hydropower (photo by ADB).

A Sunny Boost for Viet Nam's Power

Viet Nam begins its energy transition with a 50 MW solar plant just outside the capital city.

One of Viet Nam's first utility-scale solar power plants is located 50 km outside Ho Chi Minh City, the country's most populous city. The plant began operations in mid-2019 and generates roughly 78,000 MWh of clean energy annually. This will meet the needs of an estimated 40,000 households and reduce annual CO_2 emissions by around 29,760 tons starting 2020.

Viet Nam seeks to increase the share of renewable energy, including small hydro, solar, wind, and biomass power plants, as a percentage of total forecast installed capacity to 21% by 2030, and to reduce the use of imported coal-fired electricity. To meet this target, installed solar power capacity is expected to increase to 12 GW and wind power to 6 GW by 2030. Rapidly scaling up renewable energy use will also help Viet Nam achieve its target to reduce GHG emissions by 8%–25% by 2030.

The $51.5 million project was funded through an ADB loan as well as cofinancing from commercial banks, a non-parallel loan from the ADB-administered Leading Asia's Private Infrastructure Fund, and sponsors' equity.

↓29.8K

TONS OF CO_2 EMISSIONS REDUCED ANNUALLY

Inhabitants
8.99 million

GDP per capita
$6,862

Geographic area
2,061 km²

THE CHALLENGE

Expanding Viet Nam's power supply will require significant private sector investment. With limited potential to further develop hydropower and growing public environmental concerns from coal, the development of renewable energy is crucial.

CO-BENEFITS

Economic

Electricity supplied to Viet Nam's power grid will increase the availability and reliability of clean power, thereby increasing industrial and agricultural productivity.

Social

Adding new, reliable capacity to the power grid will increase access to power in remote areas, promoting socioeconomic development and increasing access to improved infrastructure and services.

50 MW solar power plant. The plant is located about 50 km outside of Ho Chi Minh City (photo by TTC Energy Development Investment Joint Stock Company).

Solar–Battery Combination for Nauru

↓11K

TONS CO$_2$ EMISSIONS
REDUCED ANNUALLY

NAURU

Inhabitants
13,300

GDP per capita
$9,397

Geographic area
21 km²

THE CHALLENGE

Historically, Nauru has relied heavily on imported diesel for power generation, creating the risk of power outages if supply is interrupted, but they are on a journey to change that.

CO-BENEFITS

Economic

The project will help reduce the cost of levelized electricity from $0.40 to $0.30 in Nauru by reducing the reliance on expensive, imported fuel.

Social

As part of the project, Nauru Utilities Corporation employees will receive technical assistance, training, and institutional strengthening.

The world's third-smallest country is investing in PV solar panels and a battery system to reduce emissions and its dependence on imported and expensive diesel fuel.

Nauru has recently invested almost $30 million in a PV and battery energy storage combination. The project will finance a 6 MW grid-connected PV solar system together with a battery energy storage system, that will be completed in 2023 and reduce over 11,000 tCO$_2$e emissions annually.

Once complete, the solar facility will generate about 12.4 MWh of power every year. During solar generation or when the batteries are delivering power, the diesel generators will be turned off and will only turn on when required. The project will provide reliable, affordable, secure, and renewable energy, reducing Nauru's dependency on diesel, and boosting the amount of electricity generated from renewable sources from 3% to 47%.

The project will directly support the target of 50% renewable energy for Nauru under the Nauru Energy Road Map 2018–2020, and assist the country in achieving its nationally determined contributions under the United Nations Framework Convention on Climate Change.

The project was funded through an ADB grant of $22 million and government counterpart financing of $4.98 million.

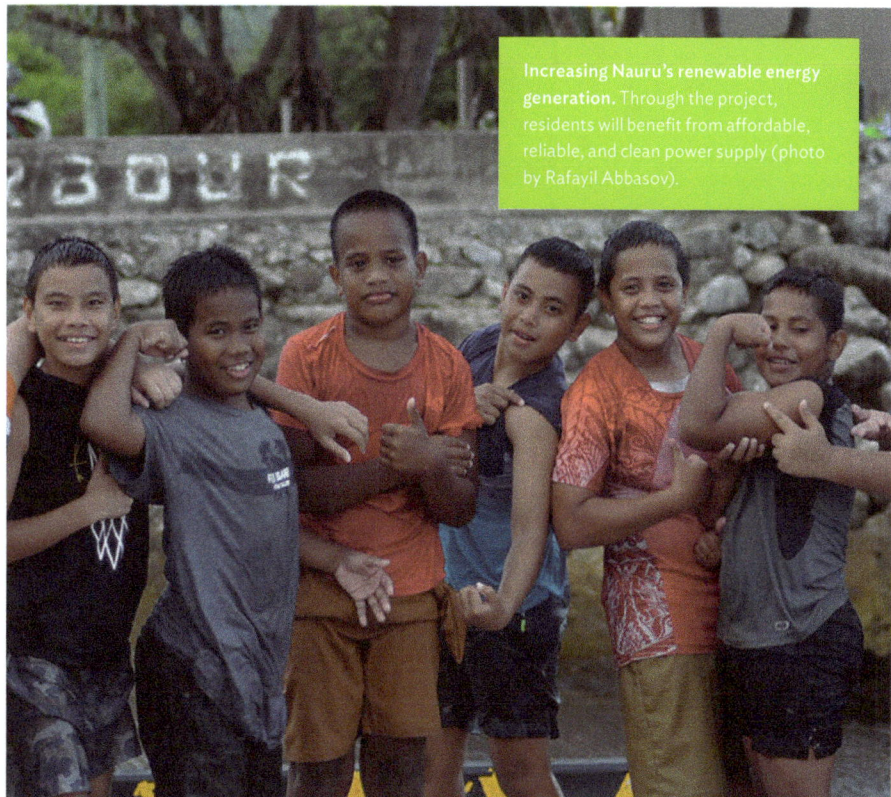

Increasing Nauru's renewable energy generation. Through the project, residents will benefit from affordable, reliable, and clean power supply (photo by Rafayil Abbasov).

Hydropower for Vanuatu's Population

Two of Vanuatu's islands are expanding the reach of the electricity grid and will connect an additional 1,050 households.

Espiritu Santo and Malekula are two of the largest islands of Vanuatu. Hydropower has been identified as the most cost-effective option for baseload power on the second-largest island Malekula where the project will construct a 400 kW hydropower plant. Although relatively small compared with energy infrastructure in more developed countries, this important renewable energy resource will supply an estimated 90% of power for the households connected to the Malekula grid.

Replacing kerosene lighting with a cheaper form of energy is expected to improve household expenditure levels. Moreover, improved electricity access is likely to also improve children's education and reduce health risks associated with burning kerosene indoors.

ADB provided a loan of $2.5 million as well as a $9 million grant, and the government cofinanced the remaining $3.1 million.

↑90%

OF POWER ON MALEKULA ISLAND WILL BE MET THROUGH HYDROPOWER

VANUATU

Inhabitants
290,800

GDP per capita
$2,993

Geographic area
12,190 km²

THE CHALLENGE

With low levels of connection to the grid, imported kerosene is the fuel of choice for indoor lighting and fuel, and diesel generators mostly power off-grid generators.

CO-BENEFITS

Economic

Fuel savings are estimated to come to $9.4 million over the economic lifetime of the project as less kerosene will have to be purchased.

Social

Improved access to electricity in Vanuatu can free up household expenditure for improved education and also means that children have better access to electricity for light to complete homework.

Health

Burning kerosene indoors can create health risks from poor indoor air quality. Reducing the reliance on this fuel indoors will therefore help improve the health of citizens.

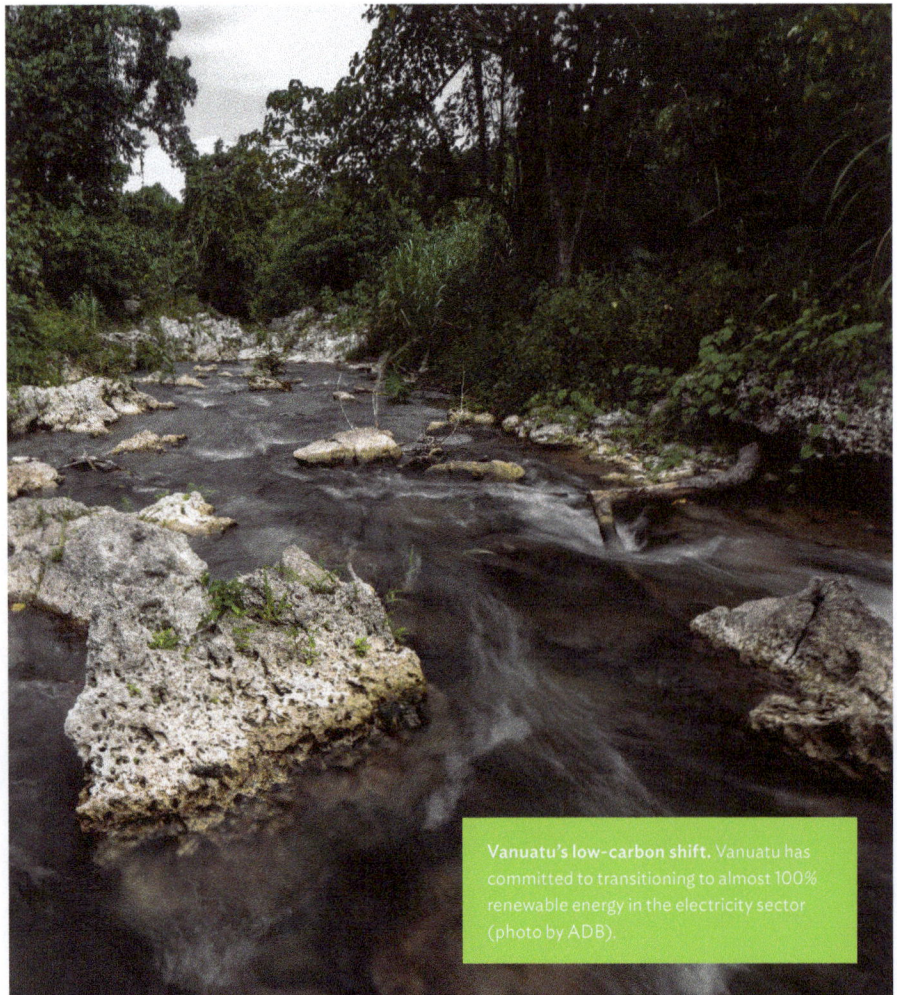

Vanuatu's low-carbon shift. Vanuatu has committed to transitioning to almost 100% renewable energy in the electricity sector (photo by ADB).

Wider reach for clean power generation. The locals of Malekula Island are the future beneficiaries of the Energy Access Project. The project will assist Vanuatu to install hydropower generation to replace diesel generation in Malekula and will extend the distribution grid in both Malekula and Espiritu Santo (photo by ADB).

Energizing Kazakhstan's South with Solar

↓141K

TONS OF CO_2 EMISSIONS
REDUCED EVERY YEAR

Inhabitants
Kyzylorda: 312,861
Shu: 37,234

GDP per capita
Kyzylorda: $2,129
Shu: $1,582

Geographic area
Kyzylorda: 240 km²
Shu: 23 km²

THE CHALLENGE

There is a significant electricity supply shortage, particularly in Kazakhstan's southern region, where the electricity sector has a reliance on outdated coal-fired plants.

CO-BENEFITS

Economic

The project will indirectly contribute to economic growth, further poverty reduction, job creation, and improvement of electricity supply in Kazakhstan.

Social

Local leaders, communities, and government units were engaged prior to the project development, and job opportunities were provided for local workers.

Kazakhstan has added two new solar power plants with a total installed capacity of 150 MW of renewable power to the energy mix as the country aims to increase the share of renewables in the electricity.

Over 1,500 PV solar panels have been installed 30 km east of the Kazakh city of Kyzylorda and over 400,000 panels close to the town of Shu. Together they have added 150 MW of clean energy to Kazakhstan's grid as part of the country's effort to further boost renewables in a nation blessed with high solar and wind potential.

Both plants lie in an area that receives more sunshine and less rain than most, and where average summer temperatures are consistently above 30°C. It is estimated that together, these projects reduce an estimated 141,000 tCO_2e emissions annually, helping the country toward the goal of making a 15% reduction in carbon emissions below 1990 levels by 2030. Power from each project is being sold under 15-year power purchase agreements.

The 100 MW plant just outside of Shu town is utilizing single-axis tracking technology, so the panels rotate throughout the day to optimize the capture of solar energy. It is the first time that such technology has been implemented in Kazakhstan.

ADB has provided a loan of up to $ 12 million as part of the $70 million project.

Boosting Kazakhstan's renewable energy generation. The solar projects are helping the country fulfill its Paris Agreement commitment of 15% emissions reductions by 2030 (photo by Baikonyr Solar Limited Liability Company).

Changsha Tests the Water with River Heating Technology

↓25K

TONS OF CO$_2$ EMISSIONS
REDUCED PER YEAR

Inhabitants
8.39 million

GDP per capita
$19,700

Geographic area
11,800 km²

THE CHALLENGE

Most of Hunan's central heating system relies on traditional heating mechanisms, including electricity and gas boilers, which release greater amounts of GHGs than cleaner alternatives.

CO-BENEFITS

Economic

The project will reduce operating costs by 40%–50%, resulting in significant energy cost savings for both residential and commercial buildings.

Environmental

The water source heat pump energy system results in no pollution, smoke, wastewater, or exhaust gas, which will improve local environmental conditions when compared to traditional heating.

Social

The central heating system will feature 24-hour uninterrupted service, supplying residents with a cost-effective and reliable heating source.

A new distributed heating project in Changsha uses river water as an energy source, cutting down on both operating costs and GHG emissions.

Two smart energy centers in the most populous city of Hunan Province will adopt river water source heat pump (RWSHP) technology, which will use water from the Xiangjiang River as an energy source for central heating.

These systems take advantage of temperature differences between the river and ambient air in both summer and winter; they will extract water as a cooling mechanism for air conditioning systems in the summer, and extract heat energy to transfer to building heating in the winter.

The energy centers, located in the new districts of Binjiang and Xiangjiang, will allow communities within a 2 km radius to connect to the heating system. Overall, around 320,000 square meters of commercial and residential buildings will be serviced by this sustainable energy system.

Compared to traditional systems, RWSHP technology will result in energy savings of around 20% in the summer and 40% in the winter. This will save the equivalent of 5,064 tons of standard coal and reduce emissions by 12,622 tCO$_2$e per year in each energy center.

A cleaner alternative for the locals. Changsha has more than 700 residential and 50 commercial users connected to regional central heating services (photo by Changsha Ecological Environment Bureau).

Carbon Finance and Partnership

→ Prohibitive costs and limited access to capital are some of the barriers to implementing renewable energy and low-carbon technologies. Through strong collaboration and innovative financial strategies, dedicated green funding can support the long-term financing of climate projects that would otherwise not be feasible.

In pursuit of climate solutions. Despite contributing a tiny fraction of total emissions globally, Maldives is taking action to move toward a sustainable energy system (photo by ADB).

Green Bonds Finance Thailand's Largest Wind Farm

↓200K

TONS OF CO$_2$ EMISSIONS REDUCED ANNUALLY

Inhabitants
1.14 million

GDP per capita
$1,857

Geographic area
12,778 km^2

THE CHALLENGE

Renewable sources contribute around 10% of the Thai energy production, emphasizing the importance of green bonds in financing. With its first green bond issuance, Energy Absolute seeks to demonstrate the potential of green bonds for clean energy innovation and contributing to climate targets.

CO-BENEFITS

Economic

Green bonds can help issuers attract low-cost capital from investors with an interest in sustainability and climate change.

Environmental

The proceeds from the green bonds are used to fund, in part or in full, new or existing projects that deliver environmental or climate-related benefits.

The importance of green bonds for financing the green transition has been highlighted by the financing of Thailand's largest wind farm.

The 260 MW Hanuman Wind Farm is the largest onshore wind farm in Thailand and is expected to reduce the country's carbon emissions by 200,000 tCO$_2$e annually. The green bonds issued for the wind farm, the first for a wind power project in Thailand, represent the country's third Climate Bonds Standard-certified climate bond issuance and the second by a Thai energy company. The issuance will support the long-term financing of Energy Absolute, one of the largest renewable energy companies in Thailand, through refinancing of short-term supplier credit that was used to help construct the wind farm.

The green bond issuance for the Hanuman Wind Farm contributes to the evolution of the green bond market while supporting the clean energy ambitions of Thailand's Power Development Plan. It is anticipated that the attractiveness of the green bond market in Thailand, which has the largest bond market among Association of Southeast Asian Nations markets, is only expected to increase.

For the project, ADB provided a Thai baht loan equivalent to $97.4 million and certified the entire green bond issuance, which attracted cofinancing of $227.3 million equivalent.

National climate targets. Thailand's Power Development Plan aims to increase the portion of renewables to 15%–20% in the energy mix by 2036 (photo by Energy Absolute).

Affordable Loans for Greener Houses

A new green loan program has been initiated in Ulaanbaatar to assist citizens with purchasing environment-friendly houses and for energy efficiency improvements.

In Mongolia's capital city, three of the country's commercial banks have collaborated to begin a new green loan program. The green loans can be used by citizens to cover purchasing of household items such as electric heaters, and building retrofitting materials and water-saving toilets in an effort to reduce air and water pollution.

As well as the green loans, residential housing loans are being offered to residents. They can choose among 10 different blueprints of single-house dwellings that meet the criteria of the loans such as size and energy efficiency standards. This pilot loan program will begin with accepting 60 applications and if proven effective, the financing will expand into the *ger* area, where the main source of Ulaanbaatar's air pollution comes from.

These initiatives are helping Mongolian banks to gradually build their green lending portfolios. As of 2019, over 2,700 green loans equal to over $60 million were financed by member banks of the Mongolian Sustainable Finance Association.

↑$60

MILLION OF GREEN LOANS HAVE BEEN FINANCED

Inhabitants
1.54 million

GDP per capita
$5,653

Geographic area
4,704 km²

THE CHALLENGE

Mongolia's capital city needs to reduce air and water pollution, and to drive these sustainability measures, three of the country's commercial banks have collaborated to begin a new green loan program.

CO-BENEFITS

Economic

The loans are subsidized by the National Committee for Environmental Pollution Reduction and are offered to citizens at half of the normal interest rate.

Environmental

The loans have financed initiatives such as renewable energy projects, water conservation, air pollution reduction, and biodiversity protection.

Green loans for green homes. The green loans assist citizens with purchasing environment-friendly houses and incorporating energy efficiency improvements (photo by ADB).

The Island Nation POISED to Embrace Solar

Maldives has begun gearing up for large-scale renewable energy production, with a project named Preparing Outer Islands for Sustainable Energy Development (POISED).

The POISED project began in 2015 and to date has helped install more than 10.5 MW of solar photovoltaic and 5.6 MWh of battery storage in Maldives' outer islands. After 25 years, the project aims to be able to reduce around 40,000 tCO$_2$e emissions per year.

On the island of Addu, the second-largest inhabited island, a 0.5 MWh lithium-ion battery energy storage system was installed with high-speed charge and discharge features and an advanced energy management system. The project is expected to contribute to increasing solar PV penetration capacity of the system from 33% to 54% and increase grid stability.

The POISED project already has installed solar PV–battery–diesel hybrid systems in over 70 outer islands, including grid upgrades, energy management systems, and Supervisory Control and Data Acquisition systems to monitor the outer islands' power system from Malé.

The project is funded from a range of sources and is receiving $55 million in ADB-administered grants—$38 million from the Asian Development Fund, $12 million from the Strategic Climate Fund, and $5 million from the Japan Fund for the Joint Crediting Mechanism. The project also has $50 million cofinancing from the European Investment Bank.

↓40K

TONS OF CO$_2$ EMISSIONS REDUCED ANNUALLY BY 2045

Inhabitants
533,900

GDP per capita
$10,791

Geographic area
300 km^2

THE CHALLENGE

The widely scattered nature of the islands together with their small size and demand make generation costs higher and create a challenge for bringing renewable installations to scale and to attract the private sector.

CO-BENEFITS

Social

Improved access to energy is expected to improve levels of education, reduce poverty, and strengthen the sense of community. There will also be job opportunities for the island residents.

Economic

Improved access to energy and lower cost of electricity are expected to improve the overall economic output and productivity of the country.

Improving clean energy access. The project targets to reach all 160 islands of Maldives through renewable power installations (photo by ADB).

Financial Support Promotes Low-Carbon Upgrades

↓**2.9K**

TONS OF CO_2 EMISSIONS
REDUCED ANNUALLY

Inhabitants
1.54 million

GDP per capita
$5,653

Geographic area
4,704 km²

THE CHALLENGE

Due to the extremely cold climate, Ulaanbaatar has a very long heating period, typically from mid-September to mid-May. This results in a huge demand for energy, where coal is the main fuel.

CO-BENEFITS

Economic

Smart building design and building energy efficiency will reduce heat and energy requirements for the building. These savings can be used for other services provided by the hospital and health centers.

Health

The program will contribute to poverty reduction by improving access to quality and affordable health services in Ulaanbaatar *ger* areas.

An upgrade to a hospital that serves some of the most disadvantaged areas in Mongolia's capital is receiving special financial support to adopt low-carbon technologies.

The Khan Uul district hospital in Ulaanbaatar serves vulnerable populations living in the peri-urban areas of Ulaanbaatar city, which has one of the world's worst air pollution levels. The project will incorporate low-carbon technologies such as energy-efficient ventilation systems, effective window insulation, ground-source heat pumps, and solar power.

Ground-source heat pumps in three family health centers will replace the heat supply from electric heaters powered by coal-fired power plants from mid-September through to mid-May in the coldest capital in the world. These technologies will contribute to the reduction of energy use from coal-fired power plants and help lower air pollution levels in Ulaanbaatar.

The project is financed with $3.48 million in grant support from the Japan Fund for the Joint Crediting Mechanism, which is one of ADB's trust funds that offers grants to incentivize the adoption of advanced low-carbon technologies. The grant also includes strategic capacity building to ensure the sustainability of the investment and continued support for operation and maintenance.

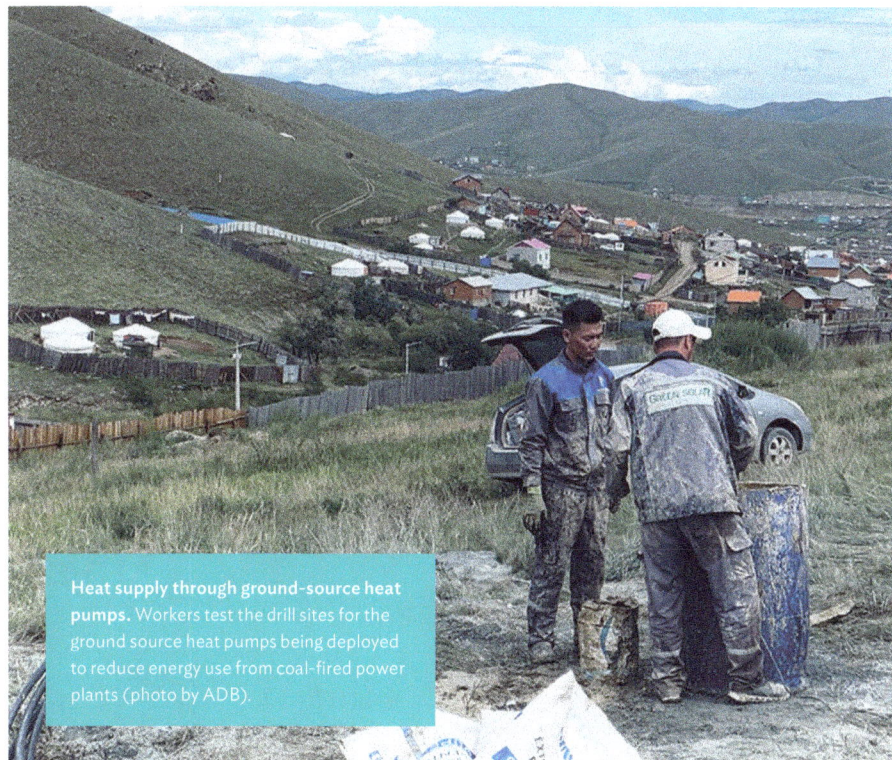

Heat supply through ground-source heat pumps. Workers test the drill sites for the ground source heat pumps being deployed to reduce energy use from coal-fired power plants (photo by ADB).

NUR-SULTAN

Connecting River Banks and Citizens
to Promote Green Transport
p. 38

NUR-SULTAN

Electrifying the Commute in Kazakhstan
p. 52

NUR-SULTAN

Block Heaters Blunt Idling Emissions
p. 44

BATUMI

Closing Batumi's Sustainable
Urban Transport Gaps
p. 40

TBILISI

Encouraging Public Transport
through Metro Face-Lift
p. 51

DUSHANBE

Leveraging the Legacy of
Tajikistan's Trolleybuses
p. 60

PESHAWAR

Gender-Inclusive Bus Corridor
to Relax Congestion
p. 47

KARACHI

Green Light for the Red
Line in Karachi
p. 46

MUMBAI

Mumbai Metro Moves the Masses
p. 54

BANGKOK

Pink and Yellow Lines
Decongest and Decarbonize
p. 55

BANGKOK

Ramping Up Thailand's Electric
Vehicle Charging Facilities
p. 41

Urban Transport and Mobility

→ Rapidly growing cities in Asia and the Pacific are tackling challenges related to urban mobility and hazardous air quality with low-carbon and sustainable transportation solutions. The road to decarbonizing urban transport often starts with incentivizing public transport, integrating clean energy vehicles, and creating more pedestrian-friendly urban design.

The PRC's First Electric Bus Network

↑50K

PASSENGERS PER DAY ON THE BRT

👥 **Inhabitants**
5.62 million

💰 **GDP per capita**
$12,288

🐂 **Geographic area**
10,942 km²

THE CHALLENGE

In recent years, the number of private cars in Jinhua has increased annually by 15%, resulting in poor air quality and congestion. In response, the municipal government is establishing an integrated, multimodal public transport system with rapid transit as its backbone.

CO-BENEFITS

👪 Social

Affordable public transport such as BRT can help improve access to public services such as health care and job opportunities for all parts of society.

♡ Health

Replacing private vehicles with BRT can reduce air pollution and improve air quality for citizens in Jinhua.

To strengthen the viable alternatives of low-carbon travel, Jinhua city in the PRC opened its first electric bus rapid transit (BRT) line, stretching 180 km through the city.

The PRC's first electric BRT first opened in 2015, providing Jinhua's 5 million citizens with a low-carbon alternative to driving—still the number one form of transport in the city.

The bus route, which has its own lane throughout the city, bypasses cars stationary in traffic and helps reduce air pollution by taking cars off the road. Jinhua's BRT has 167 buses operating across six different lines across the city, able to accommodate 50,000 passengers per day.

With the urban population growing at about 15 million people per year, almost 60% of residents now live in cities (up from one-third in 2000). With this increase comes a challenge in providing low-cost and efficient transportation. Electric BRT is one way to provide an alternative to polluting and inefficient car transport, reducing emissions from the transportation sector.

Globally, BRT has proven to be a high-capacity public transport mode that can be implemented in a short time and at a relatively low cost. The challenge lies in how BRT is perceived among citizens, as it is often considered inferior to metro systems and private vehicles.

Low-carbon public transport. Jinhua has opened the PRC's first electric BRT line, comprising 167 buses operating across six different lines in the city, and providing a low-carbon alternative to driving (photo by Yonghe Sun).

The Liuzhou Model for Mobility Electrification

The "Liuzhou Model" is a public–private partnership to encourage widespread adoption of electric vehicles throughout the city of Liuzhou which has already resulted in significant increases in electric vehicle adoption.

Through an innovative approach known as the "Liuzhou Model," the city is encouraging more and more citizens to convert to electric vehicles. This has involved assessing market demand in the city, providing test drives, promoting charging sockets, and providing special parking spaces for new energy vehicles throughout the city.

Together with the electric vehicle company SAIC-GM-Wuling Automobile, Liuzhou conducted market research to find that in 80% of city journeys, only one person is present in the car. This led to the development of a small and affordable two-seater vehicle, specifically designed for short urban transport and easy parking. A 10-month test period was also offered to citizens, after which 70% of users decided to purchase the vehicle.

Progress is being measured with metrics such as the charging station density and parking space ratio in the city. Over the course of the project to date, an additional 6,800 new parking spaces have been created for the small electric vehicles and domestic charging stations have been made accessible to owners. In the first half of 2019, Lizhou's electric car usage increased by 28% year-on-year from 2018.

↑28%

YEAR-ON-YEAR INCREASE IN ELECTRIC CAR USAGE FROM 2018

Inhabitants
4.08 million

GDP per capita
$11,151

Geographic area
18,677 km²

THE CHALLENGE

Electric vehicles are preferable to their fossil-fueled alternatives, but there still remains the challenge of integrating private vehicles in a city of over 4 million inhabitants.

CO-BENEFITS

Health

Liuzhou's air quality is improving, with particulate matter concentrations down by around 30% in 2018 compared with 2017 and the percentage of days with good air quality is increasing 16.1% year-on-year.

Economic

The project aims to stimulate the new energy vehicle industry and create more jobs within the sector. It is estimated that the new energy vehicle industry can create 66,000 job opportunities.

Electric vehicle promotion in Liuzhou. The city of Liuzhou in the PRC has a population of around 4 million and is one of the leaders in the transition to electric vehicles (photo by Fengbin Chen).

Connecting River Banks and Citizens to Promote Green Transport

↓30

MINUTES REDUCTION TO ACCESS CERTAIN AREAS

Inhabitants
1.16 million

GDP per capita
$18,743

Geographic area
810 km²

THE CHALLENGE

Nur-Sultan suffers from a combination of high levels of congestion and poor air quality. The city is partly addressing these challenges with investment in public transport and low-carbon alternatives.

CO-BENEFITS

Social

The bridge offers communal and recreational areas for citizens, and reduces commuting time to access the left side of the river, especially during rush hour.

Health

The city hopes that by encouraging walking and cycling, residents will use their cars less often, leading to improved air quality in the city.

In July 2018, a new pedestrian bridge, "Atyrau Kopiri," was opened in Kazakhstan's capital, Nur-Sultan, promoting sustainable urban mobility. It enables walking, cycling, and socializing, and connects to the right bank of the Yesil river.

The 313-meter-long Atyrau Kopiri pedestrian bridge connects the river banks of Nur-Sultan, Kazakhstan's capital formerly called Astana, and promotes zero-emission transportation modes.

The bridge's design takes inspiration from natural forms and when viewed from above, it resembles the shape of a fish, the symbol of Kazakhstan's Atyrau region on the east of the country that borders the Caspian Sea.

The bridge, connected to 56 km of additional urban bike lanes, aims to decrease traffic jams and promote green transportation such as walking and cycling. The bridge also reduces the time to access the left side of the river during rush hour by 30 minutes. Moreover, Nur-Sultan's citizens and tourists can now utilize the public space in the middle of the bridge, by listening to small-scale concerts during the summertime for example, or for jogging in a nearby park.

Atyrau Kopiri pedestrian bridge. The 313-meter pedestrian bridge connects the two banks of the river in Nur-Sultan, offering walking and cycling routes as well as recreational opportunities (photo by Na Won Kim).

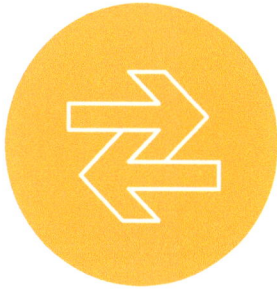

Green Buses Increase Efficiency and Decrease Emissions

↓3.15M

TONS OF CO_2 EMISSIONS
REDUCED OVER 10 YEARS

Inhabitants
9.50 million

GDP per capita
$17,986

Geographic area
11,293 km²

THE CHALLENGE

Qingdao is the largest city in Shandong province with a population of over 9 million. The city has experienced rapid development and is considered to be a major industrial center. The challenge is in matching the quality of the transport system with the pace of growth.

CO-BENEFITS

Environmental

The city intends to recycle the updated battery components from the buses.

Social

The Qingdao green bus scheme is helping alleviate the current situation of urban road congestion.

Economic

Related supporting facilities such as charging stations and batteries can provide more employment opportunities for local enterprises, promoting local economic development and technological research.

The city of Qingdao in the PRC is increasing the efficiency of its bus system to reduce emissions, ease congestion, and make public transport more attractive for the citizens.

The Green Bus System Demonstration Project in Qingdao aims to increase demand for public transport by improving the city's bus system through a combination of upgrading the fleet and adjusting the planning to prioritize buses. Over 900 new electric buses are being purchased, along with 208 charging stations and two new bus depots.

It is estimated that the bus optimization strategies and electric bus upgrades will help the city to reduce around 3.15 million tCO_2e emissions over the course of 10 years.

The lessons learned from this demonstration project may be rolled out in other large and medium-sized cities in the PRC as they seek to establish climate change adaptation measures and reduce GHG emissions.

Optimization and efficiency improvements in the public transport sector that Qingdao aims to make will help the city decouple carbon emissions from economic growth in the future.

Green bus system. In the next 10 years, 900 new electric buses will be available in Qingdao, helping to decarbonize the transport sector (photo by Jun Huang).

Closing Batumi's Sustainable Urban Transport Gaps

Amid a rising population, expanded city boundaries, and a flourishing tourism industry, the Green Cities: Integrated Sustainable Transport for the City of Batumi and the Ajara Region (ISTBAR) project seeks to decouple growth from increased carbon emissions in Georgia's second-largest city.

The Green Cities: ISTBAR project promotes sustainable urban transport solutions and supports related policy making at the city, regional, and national levels. Together with the United Nations Development Programme (UNDP), Batumi city authorities are implementing the project in the absence of national strategies and policies.

The project is pioneering the study and modeling of local transport demand and emissions, developing comprehensive plans, and piloting solutions based on these to ultimately reduce GHG emissions.

The $1.1 million Global Environmental Facility-funded project has created solutions including dedicated bus lanes, renovation and expansion of bike lanes, implementation of low-emissions buses and e-taxis, reorganization of bus networks and establishment of passenger transfer terminals, all pilots for which have been finalized in early 2020. Successful pilots will then be scaled up in Batumi and across other cities in Georgia.

↓40%

REDUCTION OF AIR POLLUTANT EMISSIONS

Inhabitants
169,100

GDP per capita
$3,852

Geographic area
200 km²

THE CHALLENGE

Batumi's transport sector is responsible for 49% of the city's GHG emissions, compared to other sectors like buildings, street lighting, municipal solid waste, etc. This is largely due to the outdated semiformal minivans that act as public transport.

CO-BENEFITS

♡ Health

The project is expected to contribute to improved air quality through a 40% reduction in vehicle pollutant emissions.

Social

Improving public transport by making them safer and giving them priority over private cars will contribute to better mobility options for women, who are more likely to use public transport than men.

Low-carbon transport. The Green Cities: ISTBAR project aims to promote public transport and reduce emissions from the transport sector in Batumi (photo by UNDP in Georgia).

Ramping Up Thailand's Electric Vehicle Charging Facilities

↓53.2K

TONS OF CO$_2$ EMISSIONS
REDUCED ANNUALLY

Inhabitants
10.16 million

GDP per capita
$16,909

Geographic area
1,569 km^2

THE CHALLENGE

The transport sector currently accounts for about 26% of the country's annual GHG emissions. Electric vehicles comprise nearly 3% of all vehicle registrations. One of the barriers to greater electric vehicle uptake is the lack of public charging infrastructure.

CO-BENEFITS

♡ Health

A cleaner environment with the reduction in air pollutant emissions can lead to improved health for the general population.

🌍 Environmental

The shift from internal combustion engine vehicles to electric vehicles will contribute to reduced local air pollution and transport sector GHG emissions.

The anticipated first climate loan in Thailand is geared toward the electrification and decarbonization of its transport sector with a countrywide electric vehicle charging network (EVCN) expansion.

Energy Absolute, the largest renewable energy company in Thailand having solar and wind power plants with a total installed capacity of 664 MW, is now expanding its operations to produce battery and electric vehicles and develop e-charging infrastructure.

Through a corporate loan to Energy Absolute, the project will support the expansion of an extensive EVCN comprising at least 3,600 charging stations across Thailand's major cities. Once established, the EVCN in Thailand will be one of the largest in Southeast Asia and will significantly reduce GHG emissions. The loan will also support the operation of the 90 MW Nakornsawan Solar and 260 MW Hanuman Wind power plants.

The loan focusing on climate change benefits, will enable the continuity and business expansion of Energy Absolute's clean energy investments. It will be certified by the Climate Bonds Initiative and will adhere to the Green Loan Principles.

For the project, ADB provided a Thai baht loan equivalent to $47.6 million.

Promoting electric mobility. More electric vehicle charging outlets like this one will be installed across Thailand (photo by Energy Absolute).

Taxis to Help Tackle Air Pollution in Mongolia

⬆156K

TAXI TRIPS COMPLETED IN
THE FIRST 7 MONTHS

Inhabitants
1.54 million

GDP per capita
$5,653

Geographic area
4,704 km²

THE CHALLENGE

In the capital city of
Ulaanbaatar, air pollution has
been described as a "child
health crisis."[3] Promoting
less-polluting forms of
transportation is one way to
reduce the risk to citizens.

CO-BENEFITS

♡ Health

The registered taxis that use
LPG produce less CO_2, carbon
monoxide and less NOx
emissions than petroleum,
reducing health risks for
citizens.

Social

An academy for taxi drivers
was created, which provide
training on topics such
as safety procedures and
professional development to
enhance their skills.

In the capital city of Ulaanbaatar, a new ride-hailing app is
helping connect citizens with taxis running on liquefied petroleum
gas (LPG), a far less-polluting alternative to petroleum and diesel.

Through a mobile ride-hailing application called UBCab, reliable taxi services are
assisting the public with access to more environment-friendly vehicles in an effort
to reduce air pollution and fill the gaps from public transport.

Of the registered taxis, 90% use LPG, which produces around 10% less CO_2 than
petroleum, around 80% less carbon monoxide, and far less nitrogen oxide (NOx)
emissions, some of the main air pollutants that can cause health risks.[2]

The UBCab application uses global positioning system and smart payment
systems to connect 70,000 registered users with taxi operators 24-7. Drivers
receive fuel discounts at select gas stations that supply LPG as an incentive for
registering their taxis in the application.

Due to the poor development of public transportation options for residents and
the city's freezing winter conditions, the taxis are helping to promote fast and
effective travel, while providing less-polluting transportation options than privately
owned vehicles. In the first 7 months of operation, the taxis had completed
156,000 successful trips.

---

[2] R. Ryskamp 2019. In-Use Emissions and Performance Testing of Propane-Fueled Engines. West Virginia University.

UBCab taxi. Environment-friendly taxis
running on LPG in Ulaanbaatar (photo
by UBCab LLC).

---

[3] United Nations Children's Fund. 2018.
Mongolia's Air Pollution Is a Child
Health Crisis. 22 February. https://
www.unicef.org/infobycountry/
media_102683.html.

Expanding Public Transport to Cut Congestion

↓25K

TONS OF CO$_2$ EMISSIONS REDUCED ANNUALLY TOWARD 2025

Inhabitants
820,000

GDP per capita
$4,784

Geographic area
130 km²

THE CHALLENGE

Vientiane has experienced record growth resulting in increased traffic congestion and the need to tackle its unsustainable transportation system through various transportation solutions, including the creation of a BRT system.

CO-BENEFITS

♡ Health

The city believes the BRT can contribute to improved health conditions for the 820,000 residents through cleaner air quality and safer roads.

👥 Social

The city believes the new transport solutions will improve conditions for vulnerable residents through safer transport options providing better job prospects.

Vientiane, the capital of the Lao People's Democratic Republic (Lao PDR), is tackling traffic congestion in its city center with the creation of a BRT system and other transportation solutions to shift transport demand away from private vehicles.

Vientiane is the fastest-growing city in the Lao PDR. The recent growth has increased the number of private vehicles, which has resulted in traffic congestion, rising incidence of road accidents, and deteriorating air quality. The number of private vehicles has risen 17% on average annually from 2000-2009, and has increased at a growth rate of more than 10% since then.

To tackle this, by 2022, the city plans to construct 12.9 km of new dedicated bus lanes and 28 enclosed stations, utilizing 45 modern BRT buses. The BRT buses will be battery electric-powered and will serve the existing 84 km network that covers the city. The project will also introduce traffic control measures, an official on-street paid parking system, and improved pedestrian facilities with universal accessibility measures.

ADB provided financing of approximately $35 million and technical assistance for the project.

Low-carbon public transport in Vientiane. The new BRT system will help shift transport demand away from private vehicles (photo by ITDP-China - image from concept design).

Block Heaters Blunt Idling Emissions

↓430K

TONS OF CO_2 EMISSIONS
REDUCED ANNUALLY

Inhabitants
1.16 million

GDP per capita
$18,743

Geographic area
810 km²

THE CHALLENGE

During a 200-day winter season in Nur-Sultan, any extra idling time can add up to increased emissions of GHGs as well as other more locally problematic air pollutants.

CO-BENEFITS

♡ Health

The engine block heaters reduce local air pollutants that adversely affect breathing air quality, especially for vulnerable groups such as the young and elderly.

📈 Economic

The engine block heaters cost $100–$250, but savings from fuel bills and repairs mean that the payback time is 1–3 years.

In the freezing winters of Nur-Sultan, having an electric heater to warm the car engine can reduce hours of unnecessary idling and tons of emissions.

With an average January temperature of –15°C, it can be difficult to start car engines in Nur-Sultan, the capital of Kazakhstan. Even if the engine does start, it can take an hour to sufficiently warm up for smooth running, causing unnecessary emissions from stationary and idling cars.

To improve this, Nur-Sultan is launching a pilot project with 100 block heaters installed in vehicles to warm the engine and interior of the car before it is needed for use—without relying on the engine running.

The block heaters are small devices installed in the cars next to the engine that generate heat from electricity, just like a kettle. The city is also installing 53 charging stations throughout the city so the heaters can be charged when needed.

Although not the final solution for Nur-Sultan's sustainable transportation, the block heating technology can help avoid such air pollutants and reduce GHG emissions substantially in the near term. Through wide deployment of block heating technology, Nur-Sultan can achieve a net reduction of around 430,000 tCO_2e emissions per year.

The $150,000 pilot project was funded through the ADB-managed Clean Technology Fund.

Block heater installation in Nur-Sultan.
Electric engine block heaters are being tested in Nur-Sultan to reduce emissions and fuel consumption from idling during Kazakh winters (photo by Andrey Terekhov).

Green Light for the Red Line in Karachi

75K

TONS OF CO$_2$ EMISSIONS
REDUCED PER YEAR

Inhabitants
15.40 million

GDP per capita
$6,000

Geographic area
3,780 km²

THE CHALLENGE

Karachi is one of the largest cities in the world without a formal public transport and mass transit system, the city must address this with a sustainable public transport network.

CO-BENEFITS

Health

The transformation will improve quality of life, reduce emissions and improve air quality, benefiting 1.5 million people, which is 10% of Karachi's population.

Social

The city believes that the poor, women, children, elderly, and people with disabilities would benefit most from the system due to its safety features and accessibility.

The new 26.6-kilometer BRT Red Line corridor being constructed in Karachi is expected to reduce emissions and improve public health in one of the world's most densely populated cities.

The Red Line BRT system, expected to be completed in 2023, will improve the mixed-traffic roadway through the inclusion of additional lanes, parking facilities, landscaped green areas, bicycle lanes, improved sidewalks, and energy-efficient street lights. The BRT fleet will also be upgraded with compressed natural gas (CNG) hybrid buses.

The new BRT line will be complemented by plans to introduce a waste–to–fuel scheme that will produce biomethane from cattle waste as fuel for the BRT buses. Over 6,000 tons of cattle waste are produced daily in Karachi's Cattle Colony and then discarded into Karachi Bay.

The new Red Line BRT project will provide city dwellers with alternative modes of transportation that the city has historically lacked. It is estimated that approximately 40% of all trips are still being made on foot.

The project was funded through a $235 million loan from ADB as well as with contributions from the Asian Infrastructure Investment Bank, Agence Française de Développement, and the ADB-administered Green Climate Fund.

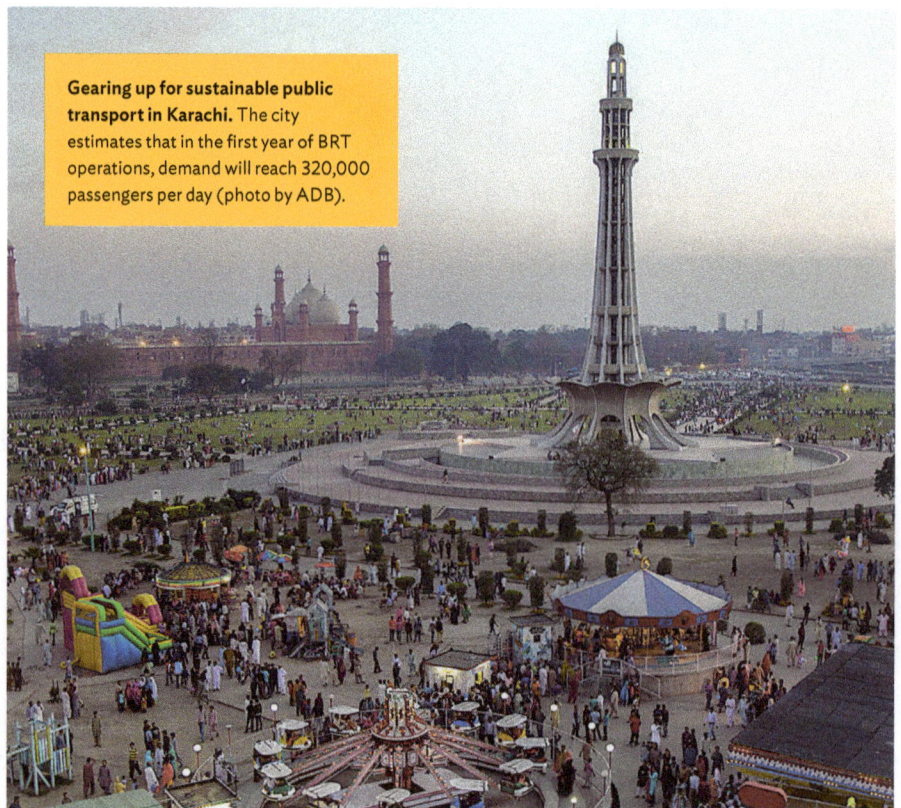

Gearing up for sustainable public transport in Karachi. The city estimates that in the first year of BRT operations, demand will reach 320,000 passengers per day (photo by ADB).

Gender-Inclusive Bus Corridor to Relax Congestion

↑500K

PEOPLE ARE EXPECTED TO BENEFIT FROM THE BRT CORRIDOR

Inhabitants
1.97 million

GDP per capita
$1,580

Geographic area
215 km²

THE CHALLENGE

Peshawar has one of the country's most congested transport systems, which affects traffic congestion and raises the risk of harassment for women. The city hopes that its new safe and inclusive BRT corridor will address these issues.

CO-BENEFITS

Social

The project hopes to encourage women's meaningful participation by ensuring that 10% of BRT operations employees are women.

Health

By encouraging commuters to switch to public transportation from private means, air quality can be improved due to the decrease in air pollutants.

Environmental

A pilot program creating bicycle parking will be incorporated to encourage commuters to include bicycles as their first- and last-mile connectivity to the BRT.

In an effort to improve air quality, reduce emissions, and relax traffic congestion, Peshawar is constructing a 26-kilometer east–west BRT corridor with gender-inclusive initiatives at the core.

Peshawar is Pakistan's sixth-largest city with 1.97 million citizens, but has one of the country's most congested transport systems. Getting from the residential district of Chamkani on the east of the city, through the center, and to Hayatabad can take well over an hour in rush hour, with average speeds of just 11 km per hour.

One way to reduce congested roads is to improve public transport options, therefore the city is installing 26 km of BRT lanes, with 31 stations, park-and-ride facilities, bicycle lanes, and green areas helping to improve air quality and reduce traffic congestion. The buses will be diesel–electric hybrids, allowing lower emissions and cleaner air compared with individual car transport.

The project will also establish universal access and safety features in all stations, including enhanced lighting, closed-circuit television monitoring, segregated areas for women, and staff trained to deal with harassment incidents.

The project was funded through a $335 million loan from ADB as well as with contributions from the Agence Française de Développement.

Peshawar's new BRT corridor. Partly commissioned in August 2020, the BRT's construction is expected to be fully completed in 2021 (photo by ADB).

Three Transport Initiatives Boost Livability and Cut Emissions

↓263K

TONS OF CO_2 EMISSIONS REDUCED EVERY YEAR

Inhabitants
2.88 million

GDP per capita
$11,371

Geographic area
5,015 km²

THE CHALLENGE

Xiangtan is currently more suited to transport by car or motorbike, and roads often prioritize private vehicles.

CO-BENEFITS

Environmental

Ecosystem-based adaptation measures will reduce pollution in downstream areas by filtering pollutants and improving runoff quality.

Social

New pedestrian pathways will accommodate wheelchair users by making sure they are wide and free from obstacles, improving accessibility for all.

Health

With better air quality, nonmotorized transport options, and safer roads, residents will be more active and thus decrease their risk for certain diseases.

Xiangtan plans to contribute to a modal shift from private to low-carbon, public alternatives through improvements to pedestrian pathways and bicycle lanes citywide, demonstrating the different functions of the road.

Three initiatives are being implemented in Xiangtan that will help boost public transportation in the city, decrease congestion, and reduce emissions by improving pedestrian, cycling, and public transportation infrastructure. The incorporation of ecosystem-based adaptation measures will improve drought, flood resilience, and air quality, while also transforming the street into a pleasant space for relaxing and resting.

The first two projects focus on walking and cycling in the city center, with improvements to 63 km of cycle lanes and 69 km of pedestrian walkways. These routes will be safely separated from roads and seamlessly connected at intersections and crossings. The pedestrian walkways will have no barriers, be free of parked vehicles, and contain smooth ramps to make walking easy for all.

The third initiative focuses on transforming the six-lane Fuxing Middle Road, which is located in a flood-prone zone. Various initiatives such as a bus priority lane, improved pedestrian pathways and bicycle lanes, will seek to transform Fuxing Middle Road into a multifunctional street. The reconstruction of a large landscaping area into a street forest will encourage walking and relaxing, and the inclusion of water-pervious street parking spaces will reduce runoff. Ecosystem-based adaptation measures like tree planting, rain gardens, and subsurface water retention boxes, will also aim to boost livability in the city.

An ADB loan has been provided to help cover the $32 million cost for all three projects.

Establishing multifunctional streets. In many areas, pedestrian and cycling lanes are not clearly marked. With ecosystem-based adaptation measures, improvements will make Xiangtan's streets multifunctional and resilient (photo by Royal HaskoningDHV).

Safe and inclusive streets. Xiangtan aims to make the streets more cycle-friendly and improve pedestrian accessibility by removing barriers (photo by Shenzhen Urban Transport Planning Center).

Improving Multimodal Transport in Xiangtan

The city of Xiangtan aims to ease the transfer between public transport options, particularly around train stations, to further promote low-carbon transport and reduce the need for private vehicles in the city.

The Xiangtan Railway Station is a large and historic station, with high-speed rail connections to other large cities such as Changsha and Zhuzhou. However, when travelers depart from the station, they are faced by a challenge if they wish to reach their final destination in Xiangtan by public transport, with no obvious options other than a taxi or car.

By investing around $2.8 million, including ADB's loan of $1.6 million, in making the station a multimodal hub, Xiangtan hopes to reduce the reliance on private vehicles and encourage the mode of public transport to reduce emissions and congestion and make the city more livable.

In this initiative, the traffic flow will be altered around Xiangtan Station and the smaller Bantang Intercity Railway Station, so that a sheltered bus stop will be close to the entrance, allowing people to switch between bus and train link more easily. The taxi queue will be relocated to a more orderly and safe location, making the bus stop more accessible.

This modification is a part of ADB's $150 million loan provided to the Xiangtan Low-Carbon Transformation Project.

↑$5M

INVESTMENT IN IMPROVED PUBLIC TRANSPORTATION

Inhabitants
2.88 million

GDP per capita
$11,371

Geographic area
5,015 km²

THE CHALLENGE

Xiangtan's railway station is not designed to promote public transport connectivity, which increases the reliance on private vehicles and taxis.

CO-BENEFITS

Economic

With transfer to buses becoming more comfortable and convenient, people will save on fuel and parking fees and streets will be less congested.

Social

This low-cost solution will be the first of its kind in the PRC, serving as a model for how a commuter-friendly multimodal hub can incentivize public transport.

Health

The modal shift from personal vehicles to public transportation will produce less pollutants, improving air quality for residents.

Targeting seamless transfers in multimodal hubs. Outside Xiangtan's multimodal stations, road barriers and taxi/private car lanes hinder public transport connectivity (photo by Shenzhen Urban Transport Planning Center).

Encouraging Public Transport through Metro Face-Lift

↓38%

DECREASE IN THE METRO SYSTEM'S ENERGY USAGE

Inhabitants
1.17 million

GDP per capita
$7,757

Geographic area
504 km²

THE CHALLENGE

Despite continued efforts to perform maintenance, the Tbilisi Metro is in need of upgrading and rehabilitation to address concerns including energy efficiency, inadequate accessibility, and lack of multimodal public transfer options at interchange stations.

CO-BENEFITS

Environmental

By rehabilitating and upgrading the Tbilisi Metro, the city hopes to reduce the metro system's energy usage by 38% compared with 2016 levels.

Social

The project will benefit residents, especially the elderly and people with disabilities, by providing safer and more reliable access to the city and other activity centers.

Tbilisi's metro system is being fully rehabilitated and upgraded, to make public transport a more attractive option for residents and help reduce emissions from the transport sector.

Tbilisi, the capital of Georgia, aims to address the safety, reliability, and efficiency concerns of its metro by focusing on upgrading its key components such as cabling, escalators, ventilation system, and water drainage pumps. Tbilisi Metro currently consists of 22 stations and two lines totaling 27.1 km. Over the years, the lack of upgrades and improvements left the metro worn out and unreliable.

The modernization project of Tblisi Metro will replace power transmission cabling, rehabilitate the ventilation system, and develop a waste management plan for its operation and maintenance. The new cables will be flame-retardant, while the new ventilation fans will improve air flow in metro tunnels, thus improving air quality and regulating temperature.

By rehabilitating and upgrading the Tbilisi Metro, the city hopes that by 2021 it will increase metro passenger trips by 19% and reduce the metro system's energy usage.

ADB provided a loan of $14.3 million for the project.

Tbilisi Metro modernization. The metro, which first opened in 1966 and became the fourth metro system in the former Soviet Union, requires rehabilitation and upgrading (photo by ADB).

NUR-SULTAN, KAZAKHSTAN

Electrifying the Commute in Kazakhstan

The capital city of Nur-Sultan in Kazakhstan has upgraded part of its bus fleet by introducing 100 electric buses.

The electric buses began being deployed in Nur-Sultan at the start of 2020 and are expected to support the local transportation system by reducing interval and waiting times. The lithium-ion battery buses contribute less to air and noise pollution than previous diesel models, and are also equipped with USB ports and Wi-Fi for the enjoyment of riders.

Electrification is widely regarded as one of the best ways to decarbonize the transport sector that has long been oil-dependent. Although currently largely reliant on fossil fuel-based power, the emissions the buses are responsible for will fall as the country increases its share of renewables in the energy mix.

In a city where air pollution from transportation is a leading health hazard, the new buses will help reduce the emission of harmful NO_x particles and improve air quality for residents.

The city is also encouraging the use of private electric vehicles with 49 new electric charging stations installed throughout Nur-Sultan.

↑**100**

NEW ELECTRIC BUSES DIVERSIFY THE BUS FLEET

Inhabitants
1.16 million

GDP per capita
$18,743

Geographic area
810 km²

THE CHALLENGE

Air pollution from transportation is a leading health hazard in Nur-Sultan. To address this, the city is upgrading part of its bus fleet with electric buses to improve air quality.

CO-BENEFITS

Economic

The city believes the project will impact the local economy by creating jobs and helping local businesses located close to the charging stations.

Health

The electric buses will reduce the amount of harmful NO_x particles in the air, thus helping to improve the health of residents.

Environmental

By replacing diesel buses with lithium-ion battery buses, the city will begin decreasing its air and noise pollution.

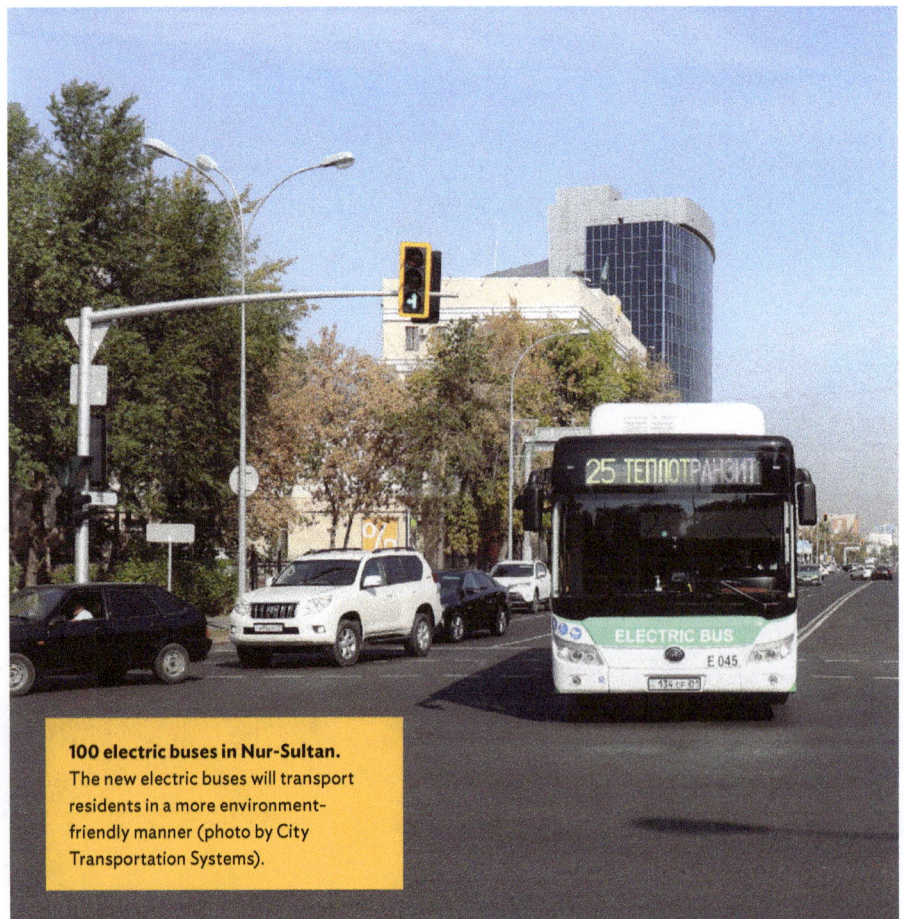

100 electric buses in Nur-Sultan. The new electric buses will transport residents in a more environment-friendly manner (photo by City Transportation Systems).

52   100 CLIMATE ACTIONS FROM CITIES IN ASIA AND THE PACIFIC

Saving Traffic, Time, and the Environment in Ha Noi

↓8.4K

TONS OF CO$_2$ EMISSIONS
REDUCED EVERY YEAR

Inhabitants
8.05 million

GDP per capita
$5,200

Geographic area
3,359 km^2

THE CHALLENGE

To encourage residents to use public transportation and reduce GHG emissions, the city of Ha Noi needs to improve accessibility, offer safe and quality services, and set an affordable ticket price.

CO-BENEFITS

Health

By choosing public transportation options over private, air quality can be improved, reducing adverse health impacts.

Social

By engaging residents through social media, the city can help change the perception of the metro and public transport and encourage residents to shift from private to public transportation.

The capital city of Ha Noi is developing an integrated public transportation system along the 12.5 km Ha Noi Metro Line 3 system with gender-sensitive and universal accessibility features.

Ha Noi is looking to improve its public transportation system by constructing a metro system network comprising eight elevated stations and four underground stations, which will improve access to five of its districts. The project will also introduce pedestrian footbridges, park-and-ride facilities, bus stops, feeder links served by 52 buses, and a real-time digital bus and train arrival information system—all providing accessibility to the Ha Noi Metro Line 3 stations.

This will become an integral part of an improved public transport system that aims to achieve increased public modal share through low-carbon transport that reduces GHG emissions.

Further, the project integrated a public opinion-based communication strategy to promote the use of the metro. The research revealed that no traffic congestion, high-speed travel, and no air pollution are the three main motivators for metro use.

The development of the integrated public transportation system is funded through an ADB loan with cofinancing from the Clean Technology Fund, Agence Française de Développement, Direction Générale du Trésor, and European Investment Bank.

Ha Noi's new Metro Line 3 system. Residents will benefit from the new metro system. The elevated section is expected to open in 2021 and the entire section afterwards (photo by Ha Noi Metropolitan Railway Management Board).

Mumbai Metro Moves the Masses

Mumbai is developing a modern and safe metro rail system in an effort to reduce GHG emissions from the transportation sector, as well as improve air quality and decongest traffic.

The Metro Rail System is currently being developed in Mumbai, which has the most crowded public transportation system in the world. The first of 12 metro lines was completed in 2014, with more due to be finalized in 2022. The new metro will employ trains using regenerative braking to reduce energy consumption by enabling the vehicle's kinetic energy to be converted back to electrical energy, as well as modern signaling systems that employ communications-based train control technology.

Safety and inclusion measures are also being introduced. Safety measures include surveillance systems, door closing and train obstacle detectors, and platforms with automatic doors. Inclusion measures include women-only carriages, mobile applications for women's security, separate women's ticket counters, reporting desks to address incidents of harassment, and priority e-ticket counters for the elderly and passengers with disabilities.

The project marks ADB's first cofinancing with the Shanghai-based New Development Bank, which will provide $260 million, while ADB will provide $926 million toward the project.

↓160K

TONS OF CO_2 EMISSIONS REDUCED ANNUALLY

Inhabitants
12 million

GDP per capita
$10,460

Geographic area
440 km²

THE CHALLENGE

Overcrowding and severe safety problems in Mumbai have led to a large number of fatalities per year and has left many commuters to rely on private transport options emphasizing the need for a new modern rail system.

CO-BENEFITS

Social

To further emphasize the metro's inclusion measures, a station staffed only by women will be established.

Health

To prevent passengers from falling onto the tracks due to overcrowding, platforms will have automatic doors that are synchronized with trains.

Anticipated ridership. Once the new metro is fully operational, millions of passengers will benefit from the system (photo by Mumbai Metropolitan Development Authority).

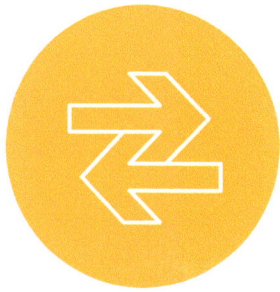

Pink and Yellow Lines Decongest and Decarbonize

↓**45K**

TONS OF CO$_2$ EMISSIONS REDUCED ANNUALLY BY 2025

Inhabitants
10.16 million

GDP per capita
$16,909

Geographic area
1,569 km²

THE CHALLENGE

Bangkok is one of the most congested cities in the world due to the congestion of its roads. By continuing to improve its MRT system, residents will keep prioritizing public transport over private.

CO-BENEFITS

♡ Health

Through the continued use of public transport, the health of residents will be improved given a reduction in the amount of harmful NO$_x$ particles in the air.

Social

The new lines will follow a universal design that will encourage use for all residents, including women, children, and the elderly and people with disabilities.

Economic

The city believes the two new lines will generate 2,000 jobs.

The Government of Thailand is addressing the congestion woes of its capital city by almost doubling the Mass Rapid Transit (MRT) lines running through the city with new pink and yellow lines.

Adding to the five existing MRT lines currently operating in Thailand's capital, the creation of the new Pink and Yellow lines will expand the options for travelers looking to escape the congested roads.

Planned for completion by the end of 2021, the new lines will expand the 100 km of existing routes to 164.5 km, with more planned for the future.

Bangkok has been reducing the amount of congestion on its roads, but it still ranks as the 11th most congested city in the world, according to navigation company TomTom's congestion index.[4] With MRT only accounting for 4% of all journeys in the city, the transportation sector is responsible for a significant proportion of GHG emissions and also contributes heavily to the often hazardous air quality.

The city hopes that the expansion of the MRT will encourage more people to switch from private vehicles to public transport, reducing CO$_2$ equivalent emissions by an estimated 45,000 tons annually by 2025 and improving air quality for citizens.

ADB provided financing of over $316 million for the project.

[4] TomTom Traffic Index 2020, www.tomtom.com/traffic-index/

Public transport promotion in Bangkok. The new lines will further encourage residents to take public transportation over private (photo by ADB).

Smart Systems and Green Vehicles Boost Public Transport

↑60%

OF TOTAL TRIPS TAKEN WITH PUBLIC TRANSPORT

Inhabitants
311,600

GDP per capita
$11,866

Geographic area
1,795 km²

THE CHALLENGE

Accelerated urbanization has strained existing networks, which consist of only 10 bus routes, and public transport ridership is marginal at 10%.

CO-BENEFITS

Environmental

The provision of clean energy public transport modes such as electric buses can reduce emissions by 30%–40%, reducing environmental pollution in the city.

Social

The project will increase environmental awareness among the youth through coding camps and hackathons related to green mobility and ITS technology.

Economic

The ITS will enable people and freight to move more easily, which can subsequently attract private sector investment and boost the local economy.

City officials aim to improve connectivity and maximize the use of public transportation in the rapidly growing city of Gui'an, with the objective of achieving 60% public transport usage by 2030.

Located in the landlocked province of Guizhou in the southwestern area of the PRC, the new city of Gui'an has a transportation system currently characterized by overcrowded road networks, high rates of traffic accidents, and low use of public transit.

The Smart Transport System Development Project is scheduled to begin construction in 2021. As part of the plan, an intelligent transport system (ITS) will be developed to improve efficiency and facilitate multimodal public transportation through a range of real-time data and management systems. This technology will be accompanied by the introduction of 200 battery-powered electric buses and a network of 21 electric charging stations. The project also will pilot autonomous vehicles and will eventually lead to mainstreaming such fleets in Gui'an's transportation area. The use of this technology complements climate change actions by enabling fuel efficiency.

The provision of clean energy transportation options alongside improved connectivity is intended to make public transport a preferred mode of travel in Gui'an, consequently decreasing congestion, reducing pollution, and making the city more livable.

The project will cost $495 million with funding coming from an ADB sovereign loan and government counterpart financing.

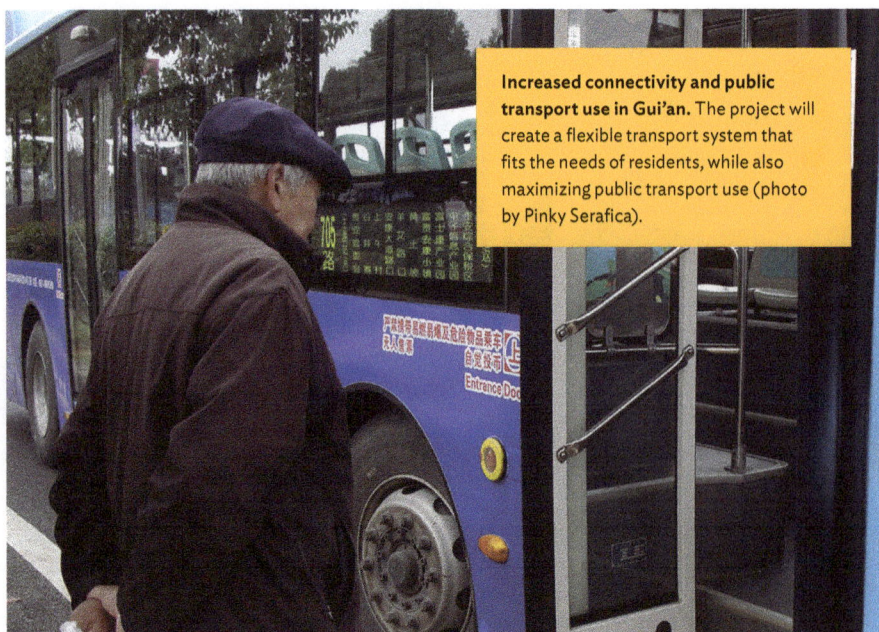

Increased connectivity and public transport use in Gui'an. The project will create a flexible transport system that fits the needs of residents, while also maximizing public transport use (photo by Pinky Serafica).

Prioritizing Public Transport to Improve Public Health

↓**60K**

TONS OF CO_2 EMISSIONS
REDUCED PER YEAR

Inhabitants
Angeles: 411,634
Mabalacat: 250,799
San Fernando: 306,659
Malolos: 252,074

GDP per capita
Angeles: $2,654*
Mabalacat: $2,654*
San Fernando: $2,654*
Malolos: $2,654*

Geographic area
Angeles: 60 km²
Mabalacat: 83 km²
San Fernando: 68 km²
Malolos: 67 km²

*regional data

THE CHALLENGE

Commuter railways operate only in Metro Manila and the Laguna province with no northbound railway linking the capital region with Central Luzon, leading to congestion and air pollution.

CO-BENEFITS

♡ Health

The new railway will provide more environment-friendly transportation options, improving air quality for residents and reducing adverse health conditions.

👥 Social

All stations will have features to provide universal access for the elderly, children, and people with disabilities.

📈 Economic

It is believed that the construction activities will create 23,900 jobs, and the operation of the railways will provide further employment opportunities for more than 1,400 people.

In an effort to reduce private vehicle dependence and pave the way for reduced transport sector emissions, the Philippines is investing in new rail lines surrounding the capital.

The Malolos-Clark Railway project supports the construction of 53 km North segment of the North-South Commuter Railway project. The line runs from Malolos to Clark International Airport. The government is developing Clark and New Clark City as a regional growth center. New Clark City will accommodate 1.2 million residents and 800,000 jobs by 2045, and will become an administrative center and education hub for the country.

The new railway line will not only reduce the reliance on private vehicles for commuters heading into Manila, but will also boost the economic opportunities for areas outside the capital city. Once complete in 2024, the new railway is expected to cut the travel time between Clark and Manila down to 1 hour from the 3 hours it can take by bus.

It will also be designed with the risk of flooding and other natural hazards in mind. The elevated alignment on viaducts will protect the line from inundation and increase safety by preventing collisions between cars and trains. Slope stabilization will also help to prevent landslides, and vegetation strategies can protect soil from severe erosion.

The new railway is the single largest project financing of ADB in the Philippines and Asia. ADB provided a loan of $2.75 billion with cofinancing from the Japan International Cooperation Agency (JICA) of $2.01 billion.

Railway for sustainability. The railway will spur economic development in the corridor and direct investments in Clark (photo by Department of Transportation, Republic of the Philippines, JICA, and the JICA Design Team).

Trolleybus Network Helps Curb Jinan's Emissions

↓250K

TONS OF CO$_2$ EMISSIONS REDUCED EVERY YEAR

Inhabitants
8.91 million

GDP per capita
$15,400

Geographic area
10,244 km^2

THE CHALLENGE

Private vehicle use is high, and the development of metro lines is not possible due to Jinan's unique underground streams and protected springs.

CO-BENEFITS

Social

The new BRT network will benefit 1.7 million people in surrounding districts, with the overall quality of public transport also being increased.

Health

By reducing particulate matter from traffic congestion, lung irritation and chronic lung diseases will be reduced and public health will be improved.

Economic

The project will create 1,500 job opportunities, and will result in lower vehicle operating costs and time savings.

A $422 million urban transport development project in Shandong's capital plans to reduce emissions and congestion in the city through the construction of a modern trolleybus network and application of travel demand management measures.

Jinan is a major administrative center in the eastern region of the PRC and is home to 1.7 million private vehicles. The number of automobiles is growing by 20% each year, and transport contributes 15% of harmful atmospheric particulate matter in the city, which is especially problematic given Jinan's status as among the 10 most polluted cities in the world.

To ensure that public transport is a more attractive option for keeping up with a steady increase in travel demand, the Jinan Public Transport Company intends to install 111 km of prioritized BRT lanes to be served by 735 electric trolleybuses.

Electric-powered trolleybuses are more energy efficient than conventional diesel buses and rail-based transport, and when integrated with BRT routes provide a high-capacity and low-carbon travel option.

Public transport ridership is expected to increase by one-third, with emissions reductions of around 250,000 tCO$_2$e each year.

ADB provided financing of $150 million for the project.

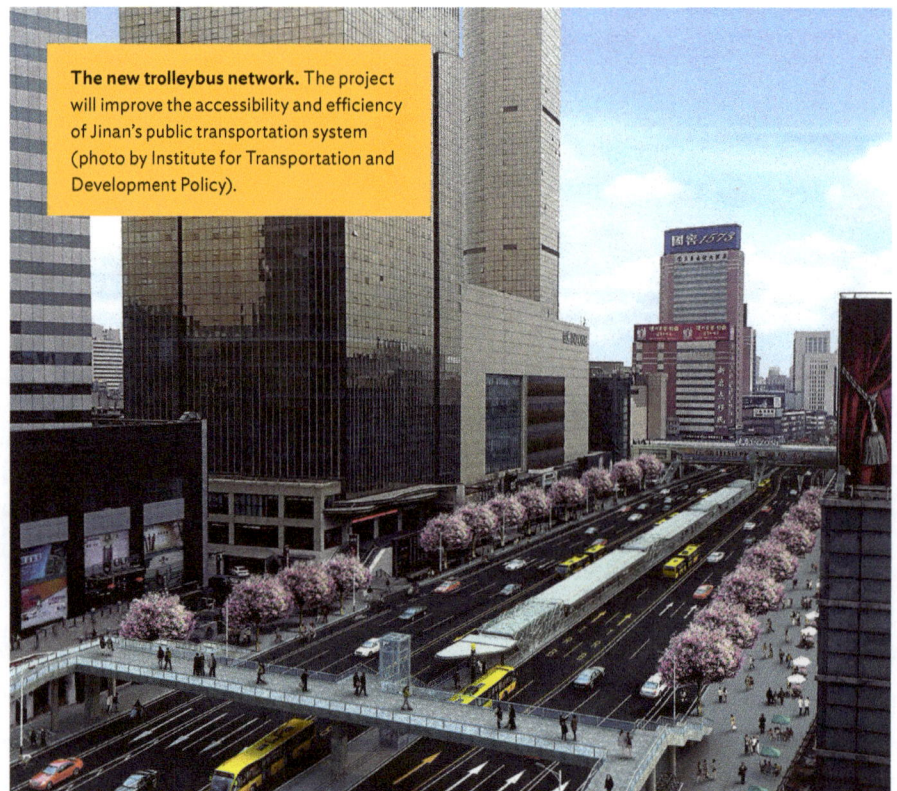

The new trolleybus network. The project will improve the accessibility and efficiency of Jinan's public transportation system (photo by Institute for Transportation and Development Policy).

Samoa Shields Harbor against Climate Change

A variety of initiatives and the construction of maritime infrastructure are underway at the Port of Apia as part of efforts to increase the resilience of the harbor.

Samoa's geographic isolation makes its only international maritime gateway in Apia extremely important. However, port operations are constrained by seasonal waves and the current breakwater is inadequate to withstand climate change and natural hazards.

The Pacific island of Samoa is planning to improve its coastal management and enhance the efficiency and safety of its port. A larger terminal area will be constructed to meet projected demand, and wave monitoring and early warning systems will be implemented to enhance safety at the port.

The project will also strengthen the existing breakwater, as current modeling has revealed that it is unable to withstand 100-year storm conditions and an expected 50-year sea level rise of 0.4 meters. Existing structures will be reinforced, the height and width will be increased, and new stormwater drainage will decrease vulnerability to flooding. Together with a multi-hazard preparedness plan, these measures seek to mitigate disruption to port operations in the aftermath of disasters. The project will also support gender-sensitive green port initiatives to promote clean and sustainable port operations and management.

The project is funded through a grant of $62.26 million from the ADB-administered Asian Development Fund and government counterpart financing of $12.77 million.

↑2.5
METERS OF EXTRA BREAKWATER HEIGHT

Inhabitants
37,391

GDP per capita
$4,323

Geographic area
124 km^2

THE CHALLENGE

As the country's only international maritime gateway, the Apia Port is critical to the economy. The seaport terminal capacity is insufficient and the severely damaged breakwater is inadequate to withstand climate change impacts and natural hazards.

CO-BENEFITS

Social

The use of locally sourced rock will benefit local contractors, and opportunities for local skilled and unskilled labor will be created.

Health

Measures designed to decrease disruptions to port operations will ensure residents are able to receive aid and emergency supplies when necessary.

Economic

Greater capacity and efficiency of port operations will increase the flow of trade and inter-island travel, and reduce the cost of doing business.

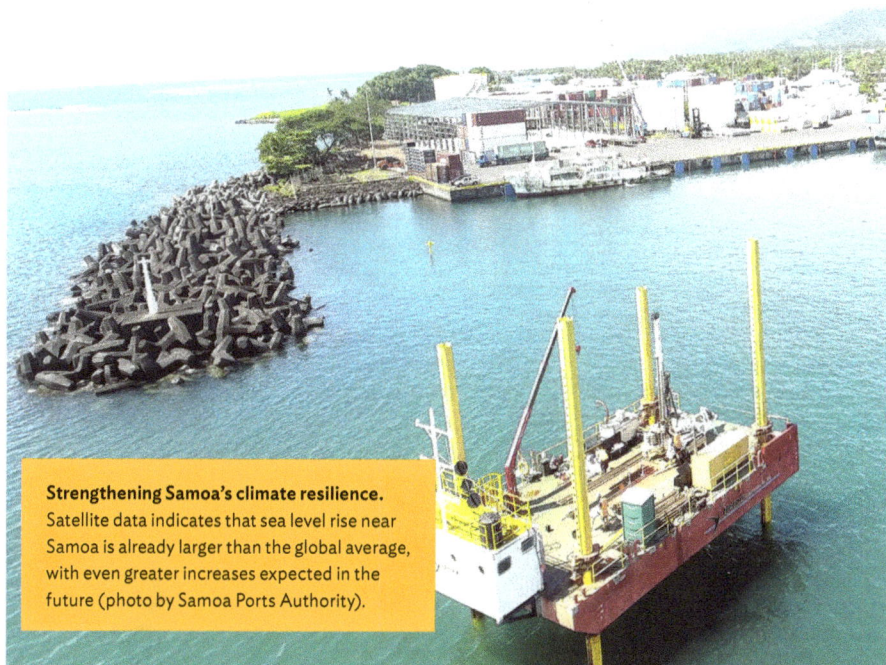

Strengthening Samoa's climate resilience. Satellite data indicates that sea level rise near Samoa is already larger than the global average, with even greater increases expected in the future (photo by Samoa Ports Authority).

Leveraging the Legacy of Tajikistan's Trolleybuses

Dating back to the 1950s, Dushanbe's antiquated trolleybus system was not keeping pace with modern life, but a refurbishment is breathing life into the system to make e-mobility popular again.

A 5-year refurbishment of the electric trolleybuses in Tajikistan's capital is bringing a transport favorite from the Soviet Union years into the 21st century. The spacious buses, connected to overhead electric cables for power, provide a low-carbon and affordable mode of transit through the city.

Trolleybuses had previously been essential for mobility in Dushanbe, providing a fossil fuel-free transportation option following the collapse of the Soviet Union and the periodic disruption of oil supplies throughout the 1990s. Trolleybuses hence thrived in the years when gasoline-powered transportation was crippled, and grew to a fleet of 250 units.

As oil supplies stabilized and competing travel modes emerged, the trolleybus system fell out of favor and were thus poorly maintained, forcing the elimination of poorly frequented routes and reduction of the fleet to 50 units.

The European Union-funded project reconstructed the old infrastructure to improve the quality and reliability of the seven-route transit system. Having facilitated 11 million passenger trips per year when it was dilapidated, city authorities predict the renewed system will boost this number significantly.

↑11M

PASSENGER TRIPS PER YEAR

Inhabitants
863,400

GDP per capita
$877

Geographic area
203 km²

THE CHALLENGE

The outdated nature of Dushanbe's trolleybus infrastructure has forced residents to rely on more heavily-polluting modes of transport, instead of incentivizing a low-carbon option powered by relatively cheap and abundant hydropower.

CO-BENEFITS

Health

Trolleybuses are considered one of the safest forms of public transport and the number of traffic accidents in Dushanbe compared with 2017 numbers decreased by almost 60%.

Environmental

The electric trolleybuses have zero tailpipe emissions, offering a low-pollution transport alternative to private vehicles and contributing to healthier air quality for citizens.

Economic

Four modern Belarusian trolleybuses were purchased as part of the project that work with energy savings of up to 45%.

Low-carbon transport for Dushanbe. Dushanbe, home to over 800,000 people, is the capital city of mountainous Tajikistan. The project is expected to boost e-mobility among locals (photo by ADB).

Daily Commute Takes New Heights

The metropolitan area surrounding Manila is aiming to attract more pedestrians and increase the use of public transportation with new elevated walkways.

A 5 km elevated pedestrian walkway along EDSA is being constructed. The EDSA Greenways project will have four common stations in Balintawak, Cubao, Guadalupe, and Taft Avenue, all along the EDSA thoroughfare. The new paths will also connect to other public transportation options to improve the area for pedestrians and encourage the switch from private to public transportation.

The construction and renovation of footbridges, and addition of elevators, will protect pedestrians from the elements and help to improve access for the elderly and people with disabilities.

By enhancing the efficiency of public transport interchange, it is anticipated that more commuters will forego driving their cars in favor of the more attractive public transportation options. There is also a plan to link the city through the ferry system.

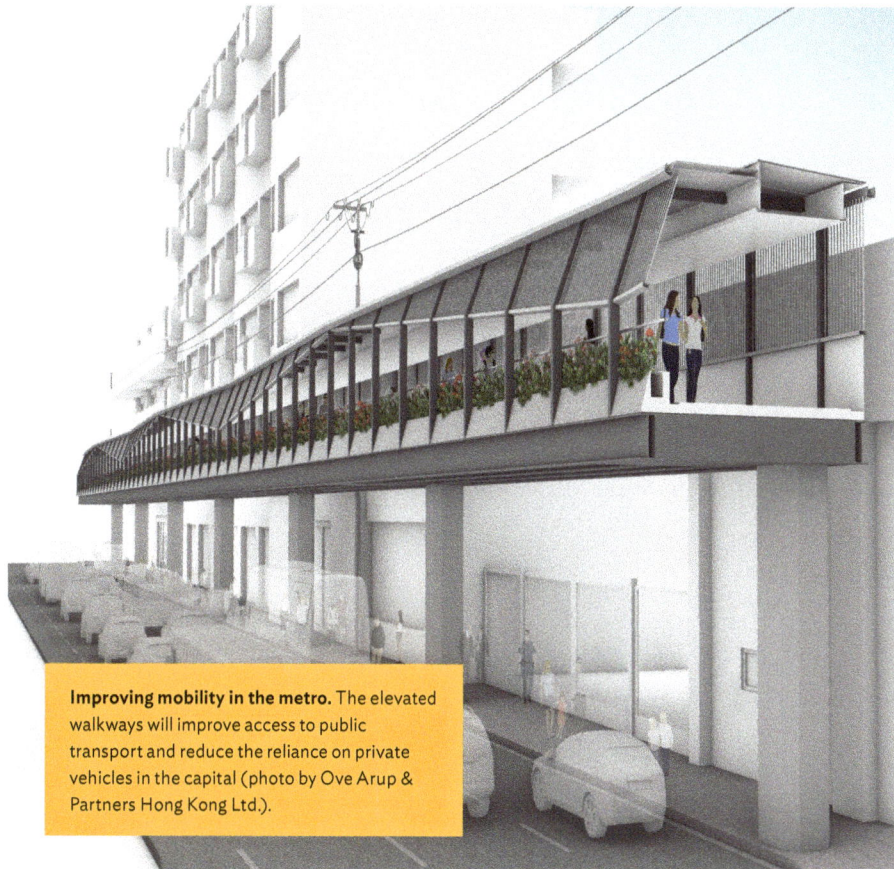

↑80%

OF TRIPS ARE TAKEN WITH PRIVATE TRANSPORT IN SOME AREAS

Inhabitants
Quezon City: 2.94 million
Makati : 582,602
Pasay: 416,522

GDP per capita
Quezon City: $9,507*
Makati: $9,507*
Pasay: $9,507*

Geographic area
Quezon City: 166 km²
Makati: 22 km²
Pasay: 14 km²

*regional data

THE CHALLENGE

In some areas, such as Quezon City, private transportation comprises over 80% of trips. By promoting integrated public transportation, the city can reduce GHG emissions from transportation and improve air quality.

CO-BENEFITS

♡ Health

The elevated walkways are designed to improve pedestrian safety and comfort, while at the same time reducing conflicts on the street level between pedestrians and motor vehicles.

Economic

Wider economic benefits are expected for local businesses as well as the greater region through congestion reduction.

Improving mobility in the metro. The elevated walkways will improve access to public transport and reduce the reliance on private vehicles in the capital (photo by Ove Arup & Partners Hong Kong Ltd.).

Land Use and Forestry

→ Land use and forestry projects can play a big role in climate change mitigation and adaptation. Trees are being planted in and around cities in Asia and the Pacific to leverage a multitude of benefits including carbon sequestration, flood protection, reduced heat stress, and improved recreational opportunities.

Billion tree tsunami. Khyber Pakhtunkhwa is in the northern region of Pakistan and is the center of large-scale reforestation efforts (photo by Mehtab Farooq via Unsplash).

KHYBER PAKHTUNKHWA, PAKISTAN

A Green Wave of Reforestation

A region of Northern Pakistan has successfully reforested an area with 1 billion trees in just 2 years, protecting vital natural ecosystems and the services they provide.

Following decades of deforestation in Pakistan, a push by the regional government saw a mass reforestation effort leading to the planting of 1 billion trees and the conservation of more than 350,000 hectares.

The so-called "billion tree tsunami" project is a part-mitigation, part-adaptation effort from Pakistan, a country that contributes less than 1% of global emissions, but is in the top 10 countries expected to be worst-hit by climate change.

The reforestation efforts have occurred in the northern province of Khyber Pakhtunkhwa, where 40% of the country's remaining forests occur. After years of deforestation, forests accounted for just 2% of Pakistan's total land area.

In preparation for the "tsunami," 13,000 private and government nurseries were set up, which contributed 40% of the new trees in the region. The remainder have come from protection and regeneration of existing lands.

↑1B

TREES PLANTED

Inhabitants
35.52 million

GDP per capita
$4,328

Geographic area
74,521 km²

THE CHALLENGE

To restore a previously deforested area and contribute to Khyber Pakhtunkhwa's Green Growth Initiative, two approaches of protected natural regeneration and planned afforestation were set in motion.

CO-BENEFITS

Economic

The economic benefits are estimated to be around $121 million for the province in terms of carbon sequestration, watershed improvement, and sustainable future forest resources.

Social

An estimated 500,000 green jobs have been created by the nurseries, which can generate incomes of around PRs12,000–PRs15,000 ($115-$140) a month for villagers.

Environmental

Large-scale reforestation is also expected to have positive effects for biodiversity conservation, as forests provide habitat for a broad range of species.

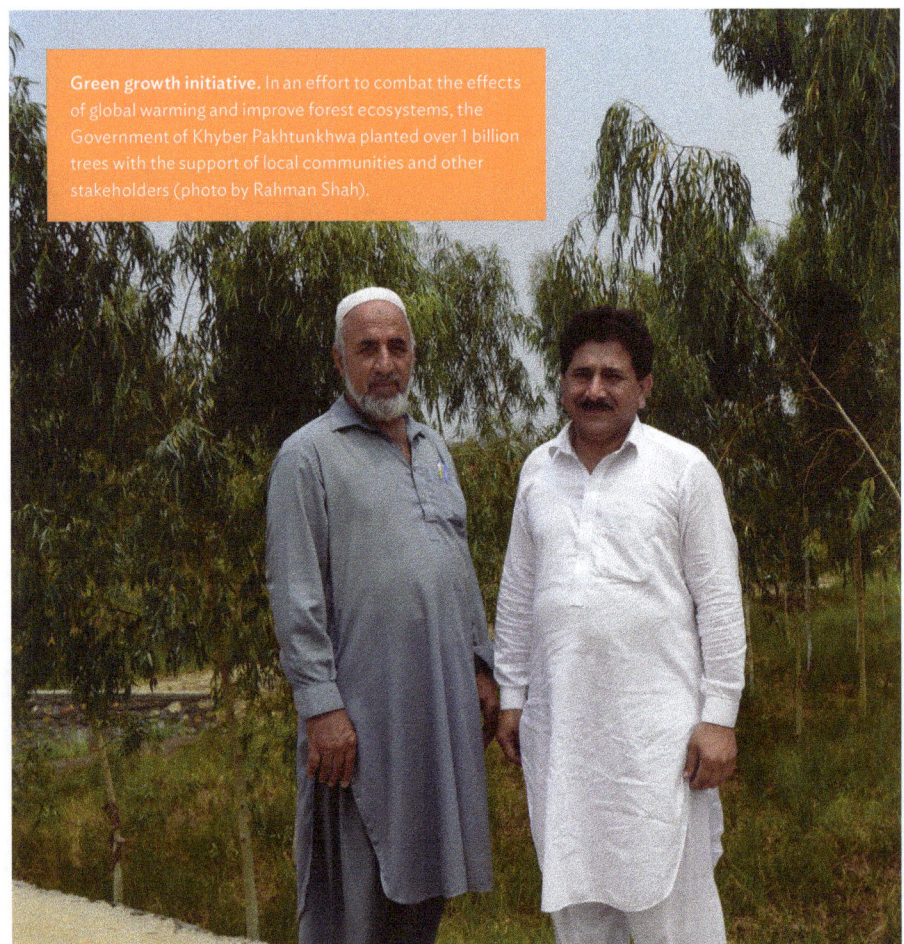

Green growth initiative. In an effort to combat the effects of global warming and improve forest ecosystems, the Government of Khyber Pakhtunkhwa planted over 1 billion trees with the support of local communities and other stakeholders (photo by Rahman Shah).

Planting Trees on the Roof of the World

The capital of Qinghai Province in the western region of the PRC has initiated an afforestation and reforestation project on the Tibetan Plateau on a huge scale.

From 2016–2018, the city of Xining introduced more than 39,000 hectares of forest within the city and in neighboring counties. In the high-altitude and semi-arid region of the PRC, this is no small challenge. By focusing efforts on barren and degraded grasslands, the project expects to see greater levels of carbon sequestration, while also improving the local environment and economy.

The degradation of land in Qinghai Province is expected to be further exacerbated by climate change, with temperatures increasing three times faster on the Tibetan Plateau than the global average.[5]

Xining is the largest city on the Tibetan Plateau and has an average elevation of 2,200 meters above sea level. Although it is more challenging and expensive to conduct afforestation efforts in higher altitudes, the project was able to select tree varieties more suited to the local climate, and organized training sessions for local farmers to spread knowledge of effective seed cultivation and forest management techniques.

---

[5] Reuters. 2018. Temperatures Significantly Rise on the PRC's Qinghai-Tibetan Plateau: State Media. 27 October.

Xining's greening efforts. The project will increase carbon sequestration and support the sustainable development of the local economy (photo by Xu Shouxiang).

↑39K

HECTARES OF FOREST INTRODUCED

Inhabitants
2.39 million

GDP per capita
$7,149

Geographic area
7,660 km²

THE CHALLENGE

Xining suffers from extremely poor air quality, soil degradation due to pressure from livestock, and vulnerability to harsh weather events.

CO-BENEFITS

Economic

During the project implementation, 35,000 jobs were created for land preparation and planting, and 150 long-term forest management jobs were established.

Health

Since trees can act as air filters by removing particulates that cause respiratory diseases, the afforestation project will improve the health of residents.

Environmental

By increasing the forest cover, the project will enhance carbon sequestration, increase biodiversity, prevent erosion, and improve water conservation.

Nur-Sultan's Trees Are Coming of Age

A 30-year long afforestation project around Kazakhstan's capital aims to create a protective green belt around the city by planting trees in the previously barren steppe.

Nur-Sultan has made a significant effort to overcome environmental challenges related to local climatic conditions, emissions, and air pollution through the creation of a forested ring around the city. More than 83,000 hectares of forest have been introduced since the late 1990s, with more than 5,000 coniferous and deciduous trees added every year.

Scientists first embarked on the large-scale afforestation project in 1998. Learning from early results, the city has seen improvements in tree survival rates. The forests are now home to a variety of animals and plants, and are being used by citizens in the summer for recreational activities.

Afforestation has been hailed as a strategy with significant global carbon sequestration potential, and although not the original goal of the project, will be an important result. The trees within and around the city will also create a number of other benefits for residents of Nur-Sultan, including protection from extreme weather events as well as a reduction in wind speeds hitting the city.

↑83K

HECTARES OF FOREST INTRODUCED

Inhabitants
1.16 million

GDP per capita
$18,743

Geographic area
810 km²

THE CHALLENGE

Kazakhstan's capital Nur-Sultan lies in the northern steppe region and is characterized by high winds, freezing winter temperatures, and baking-hot summers. These climatic factors combined with saline soils made for serious initial challenges.

CO-BENEFITS

Social

The Green Belt is a place for citizens to enjoy leisure activities such as running, cycling, and walking. Bicycles are available to rent to explore the area. The project has also created local jobs in planning, planting, and operations.

Environmental

Large-scale reforestation is also expected to have positive effects for biodiversity conservation, as forests provide habitat for a broad range of species.

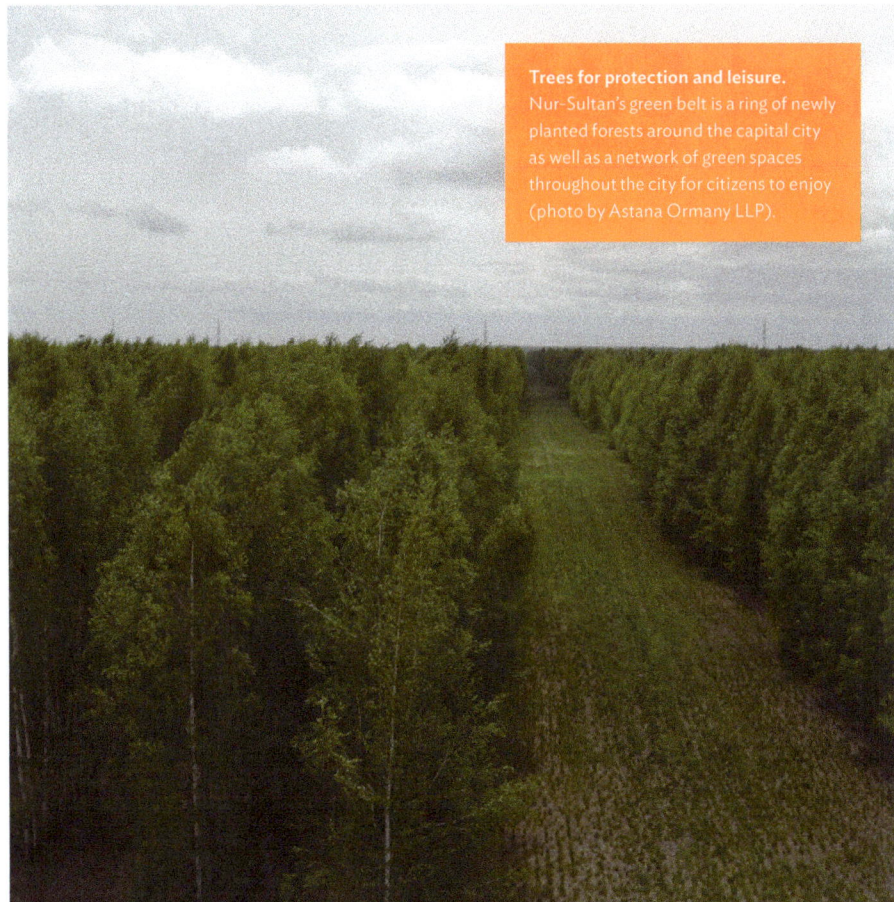

Trees for protection and leisure. Nur-Sultan's green belt is a ring of newly planted forests around the capital city as well as a network of green spaces throughout the city for citizens to enjoy (photo by Astana Ormany LLP).

Smart Cities

→ Information and communication technology (ICT) can play an important role for cities taking climate action—both in mitigation and adaptation. Cities in Asia and the Pacific are working with a number of innovative tools to record and analyze relevant data, the insights from which are being used to monitor resilience efforts, increase energy efficiencies, and implement smarter urban planning.

The making of a smart city. More than 2.8 million people live in Xiangtan, a city where digital strategies are being tried and tested to help drive sustainable development (photo by Fang Yang).

Building a City's Digital Backbone

With a range of smart city initiatives underway to help Xiangtan take action on climate change, the city is also in the process of building an overarching digital system to help manage all of them coherently.

Xiangtan is introducing a whole host of low-carbon initiatives throughout different sectors, many of which either rely on, or have been created in collaboration with, digital tools. To facilitate the management and interoperability of the different systems, the city is also creating an overarching ICT system. The new system will have an open and scalable architecture to allow further expansion and integration of sector-specific digital platforms, and will allow data sharing, verification, authentication, and security management.

The citywide ICT system in Xiangtan will integrate intelligent transport, smart bus information, building energy management, industrial energy and utility management, early flood warning, and environmental monitoring and analysis systems. The Xiangtan smart digital platforms will allow for better data gathering that would inform low-carbon behaviors and practices.

It is hoped that the digital architecture will allow more synergies across the city, combining multisector data from separate platforms. In this way, it should also allow greater collaboration among different city departments and improve the efficiency of decision making, especially when it comes to reducing disaster vulnerability.

ADB provided a loan to cover the $16 million cost for the citywide ICT platform, which is part of ADB's $150 million loan for the Xiangtan Low-Carbon Transformation Project.

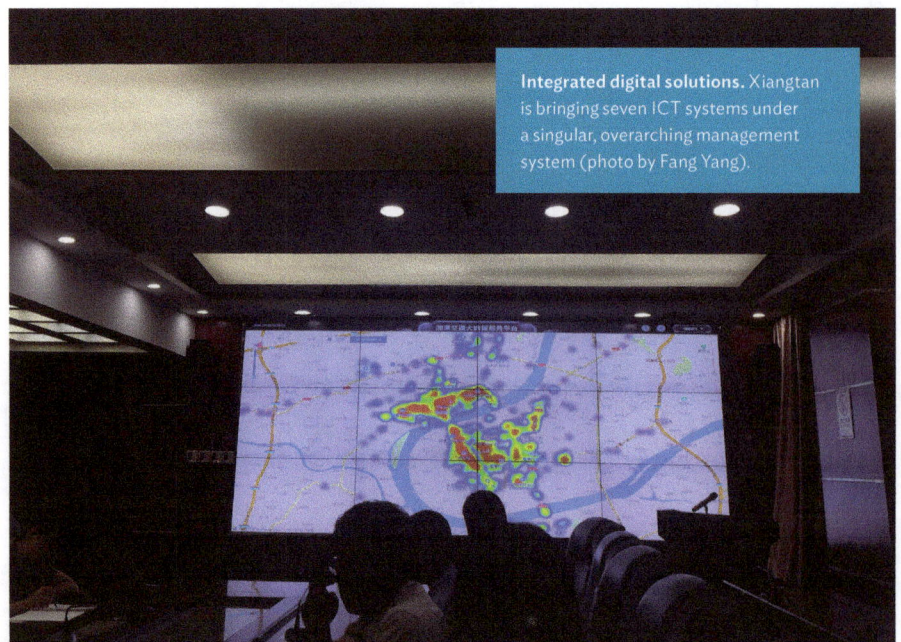

↑7

DIGITAL SYSTEMS MANAGED UNDER ONE PLATFORM

Inhabitants
2.88 million

GDP per capita
$11,371

Geographic area
5,015 km²

THE CHALLENGE

While digital solutions can hold great potential for climate action, if they are implemented in a siloed approach across different city departments, they may not be used to their maximum potential.

CO-BENEFITS

Environmental

The citywide ICT system will result in better city planning and improved disaster response to minimize environmental impact and reduced vulnerabilities.

Economic

A citywide ICT platform is a better way of managing digital information to deliver and manage better performing assets, on time and on budget.

Integrated digital solutions. Xiangtan is bringing seven ICT systems under a singular, overarching management system (photo by Fang Yang).

Energy Management System Drives Energy Efficiency

↓238K

TONS OF CO_2 EMISSIONS REDUCED ANNUALLY FROM 2030 TO 2045

Inhabitants
2.88 million

GDP per capita
$11,371

Geographic area
5,015 km²

THE CHALLENGE

Emissions from industries are currently responsible for more than 50% of total emissions from Xiangtan. This program seeks to better understand the energy usage of industry, improve efficiency by as much as 20%, and equip the city for higher levels of renewables in the future.

CO-BENEFITS

Economic

The CMEUMS is helping to optimize energy consumption and balance, saving costs for industrial companies.

Social

The CMEUMS creates a culture of collaboration among companies to control peak and non-peak loads, and optimize energy and resource efficiency at a zone level.

Xiangtan is installing a community-scale multi-energy and utility management system (CMEUMS) in the Jiuhua Economic and Technological Development Zone to monitor energy, water, and gas consumption and drive energy efficiency gains.

Xiangtan will install a CMEUMS, connecting more than 1,300 enterprises in the Jiuhua Economic and Technological Development Zone, located in the northwest of the Xiangtan urban area.

The system will monitor electricity, gas, and water consumption of the industrial plants, businesses, and buildings in the zone in order to drive the improvement of operational efficiency of each enterprise. In addition, it will enable the regulation of consumption peaks with differentiated fee structures and encourage trade among plants for wasted heat and/or gas.

Using smart technologies and the Internet of Things, increasing the CMEUMS coverage at the Jiuhua industrial zone from 2030 to 2045 is expected to result in average annual savings of 238,185 tCO_2e per year, assuming a long-term growth plan.

ADB provided a loan to cover the $4.5 million project cost, which is a part of ADB's $150 million loan to the Xiangtan Low-Carbon Transformation Project.

The Xiangtan CMEUMS. The software will be developed and installed in the Xiangtan Municipal Big Data Center and all data will be collected in and sent from there (photo by Fang Yang).

Increasing Building Efficiency through Smart Monitoring

↓98K

TONS OF CO$_2$ EMISSIONS
REDUCED EACH YEAR

Inhabitants
2.88 million

GDP per capita
$11,371

Geographic area
5,015 km^2

THE CHALLENGE

Rapid urbanization in Xiangtan has resulted in the growth of the city's urban development area and a simultaneous increase in emissions from buildings, many of which need to adopt new measures to increase energy efficiency.

CO-BENEFITS

Economic

The new monitoring systems are expected to improve energy efficiency in buildings by 10%, which translates to cost savings that can be invested elsewhere.

Health

Energy management systems will also improve the comfort of residents through air circulation and automatic climate controls.

Environmental

The project will result in energy savings of almost 24,000 MWh per year, reducing the amount of coal needed to produce energy for the power grid.

A new building energy management system will be installed to monitor electricity, water, and gas consumption in 200 public buildings in Xiangtan.

While clean and renewable sources of energy have increased in the PRC in recent years, many regions still rely on coal-powered facilities. This means that any excess power or heat leads to more emissions, and that energy efficiency measures can be an important tool for reducing building sector emissions.

Energy management systems for buildings can be the first step for municipalities to understand how and when energy is used, before taking further steps to cut emissions from energy sources and buildings.

Xiangtan is installing new building energy monitoring systems that will cover 200 public buildings, equaling around 900,000 square meters of floor space. By using smart technology and the Internet of Things, a building energy management system can manage both energy and utility systems, and works to promote demand-side energy conservation through timing schedules, occupation detection, and weather-based demand forecasts.

ADB provided a loan to cover the $5.7 million cost for the building energy system, which is part of the ADB's $150 million loan to the Xiangtan Low-Carbon Transformation Project.

Toward achieving building emissions reduction. It is expected that building energy management systems will be further installed in new and retrofitted buildings in Xiangtan from 2022 to 2045 (photo by Fang Yang).

Real-Time Urban Flood Forecasting and Warning System

↑500+

SENSORS PROVIDING
REAL-TIME DATA

PUDONG

Inhabitants
6 million

GDP per capita
$33,209

Geographic area
1,210 km²

THE CHALLENGE

Factors such as rapid
urbanization, dense river
network, relatively flat terrain,
and strong tidal influence make
Pudong vulnerable to extreme
weather conditions, storm
surges, and climate change.

CO-BENEFITS

Environmental

The flood forecasting and
warning system gives timely
notice for drainage operations,
reducing damage to surrounding
ecosystems and communities.

Social

The system allows delivery of better
water safety services, supports
flood risk assessment, information
dissemination, and strengthens
overall flood management,
including preparedness, response,
and recovery.

Economic

Early flood warning minimizes
human fatalities, injuries and
health risks, and infrastructure
damage, and reduces
costs related to post-flood
rehabilitation and rebuilding.

A real-time urban flood forecasting and warning system was built for Pudong New District in Shanghai covering 1,210 square kilometers of rapidly developing urban area east of Huangpu River.

The coastal metropolis Shanghai is susceptible to flooding due to its low-lying terrain, massive urbanization, and location downstream of the large Taihu Basin. Although local water authorities were collecting weather and hydrological data, more intelligent data technologies were needed to help make the right decisions for flood prevention and management.

A digital urban flood forecasting and early warning system was built to integrate all of the data and support water management, enabling flood risk evaluation with just a few clicks. With the help of meteorological forecast estimates, and real-time rainfall and river level data received from stations all over the Pudong New District, rapid forecasting technology allows online monitoring and simulation of flood events, providing a prediction of the scale, timing, and location of impending floods.

The early warning system improves response management of watershed and flood management agencies, allowing Pudong to optimize utilization of flood management infrastructure especially where there is limited flood carrying service capabilities.

Smart system for informed decisions.
Shanghai has implemented a real-time flood warning and forecast system, based on a range of digital inputs (photo by Ewaters Environmental Science & Technology (Shanghai) Co., Ltd).

Smart Bus Riders Experience "Green Waves" and Get Priority

↓200K

TONS OF CO$_2$ EMISSIONS REDUCED PER YEAR

Inhabitants
2.88 million

GDP per capita
$11,371

Geographic area
5,015 km^2

THE CHALLENGE

Public transport use in Xiangtan is relatively low at around 19%. By contrast, private vehicles are popular among the city's inhabitants, with ownership having increased by over 450% from 2008 to 2016.

CO-BENEFITS

♡ Health

Increased public transportation use will result in fewer pollutants, improving air quality and public health in Xiangtan.

〰 Economic

A more efficient transport system will reduce costs associated with transportation for residents, while also making commuting faster and more reliable.

👥 Social

The city aims to create a rapid, safe, and efficient movement of buses through the intersections and reduce travel time for bus users, making traveling by buses more appealing.

Xiangtan is implementing the PRC's first low-cost median bus priority lane, combined with "green wave" traffic light technology to improve efficiency of bus movement through traffic.

Physical and digital infrastructure improvements to Xiangtan's bus system are intended to improve services, incentivize public transportation, and reduce private vehicles on the road. Physically, 31 km each of median bus lanes and peak-hour curbside priority lanes will help to improve services and make commuting faster. A total of 70 new smart median bus stops and real-time information will also improve access. The network will connect Xiangtan's main transport hubs, with designs based on predicted passenger flow and future growth.

Digitally, the city's intelligent transport system (ITS) is being modified to shift the focus from cars to instead prioritize buses and pedestrians. This means that the existing system's 225 traffic lights and 1,100 traffic sensors will be adjusted to allow more efficient public transport through priority signaling and green waves for buses. Green waves are created when multiple sets of lights in a row turn green when approached by one type of vehicle. This system will be the first in the PRC with bus-designated traffic lights and a green wave system.

This project is part of the Xiangtan Low-Carbon Transformation Project, covered by a $150 million loan from the ADB.

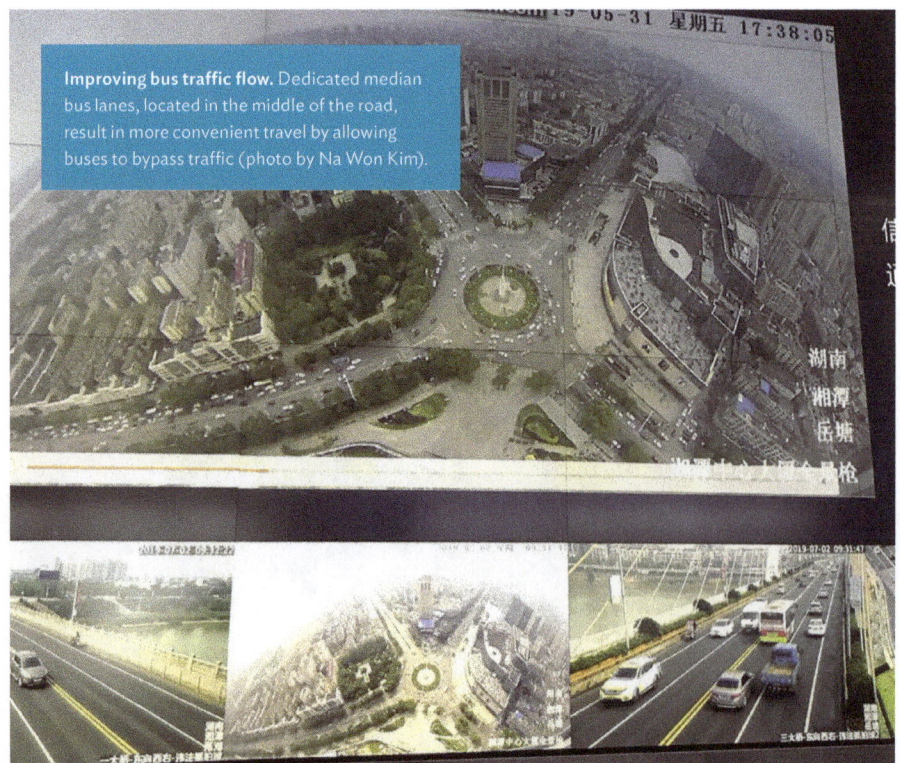

Improving bus traffic flow. Dedicated median bus lanes, located in the middle of the road, result in more convenient travel by allowing buses to bypass traffic (photo by Na Won Kim).

Flood Forecasting Enables Quick Responses and Smart Planning

↓**6K**

HECTARES EXPECTED TO SEE
REDUCED FLOODING

Inhabitants
14.68 million

GDP per capita
$3,718

Geographic area
206 km²

THE CHALLENGE

Flat terrain and insufficient drainage amplify flooding and water logging in the city, which experiences heavy rains during the monsoon season.

CO-BENEFITS

Economic

The project's cloud-based platform reduces operational costs associated with physical servers, and site-specific warnings can reduce losses for businesses.

Environmental

The time series data provided by the system can inform studies on heat stress and air quality, allowing the city to create sustainable urban plans.

Social

Early warning systems and real-time inundation data allow the city government to plan and disseminate evacuation information to vulnerable populations.

A city-level initiative that is the first of its kind in India will allow communities in Kolkata to learn from past events and act swiftly in the face of hazards through forecasts, real-time information, and a wide range of climate data.

Kolkata Municipal Corporation is working to make Kolkata a smart and resilient city by reducing urban flood risks through the introduction of a flood forecasting and early warning system.

Kolkata is one of the densest megacities in the world, and is consistently ranked in the top 10 most vulnerable cities due to its high risk of flooding under climate change projections. The new system is composed of a low-cost, high-density network of 350 sensors that will provide real-time information on flooding, air quality, heat stress, and humidity, the data from which will enable remote monitoring via the "Kflood" website.

Since most floods in Kolkata are caused by local rains, the response time is often less than 1 hour. The construction of an early warning system will benefit residents by providing timely warnings about impending natural hazards, reducing the economic loss and impact on livelihoods for vulnerable communities.

The project has been implemented under the Kolkata Environmental Improvement Investment Program supported on a grant basis by the ADB-managed Urban Climate Change Resilience Trust Fund (UCCRTF).

Data-driven solution. By including a wide range of data from sensors like this one, the new system can be utilized year-round, instead of only during monsoon season like traditional forecasting systems (photo by UCCRTF).

DUSHANBE

Dushanbe Doubles Down on
Water Security
p. 93

DHAKA AND KHULNA

Increasing Resilience in Bangladesh's
Rapidly Growing Cities
p. 85

DHAKA

Dhaka Prepares for Reduced
Freshwater Availability
p. 94

PHUENTSHOLING

Bhutan Battles Rising Waters in the
Himalayan Foothills
p. 95

KABUL

Efficient Stoves Protect Lives,
Forests, and the Climate
p. 78

TAMIL NADU

Tamil Nadu Invests in Climate-
Resilient Infrastructure
p. 88

Sustainable and Low-Carbon Communities

→ Cities in Asia and the Pacific are finding new ways to expand
the provision of urban services to cover growing populations, while
incorporating low-carbon and climate-resilient measures. Improved
infrastructure, new technologies, increased public awareness, and
nature-based solutions can all work together to improve living conditions,
while also reducing the impact of urban growth on the environment.

Efficient Stoves Protect Lives, Forests, and the Climate

The introduction of efficient cookstoves to families through Afghanistan's capital Kabul is improving health, reducing emissions, and slowing deforestation rates.

The "Efficient Cookstoves for Women in Afghanistan" project is distributing modern and efficient cookstoves to families throughout the capital Kabul, and is funded by a grant from the United Nations Development Programme (UNDP) and the Government of the Republic of Korea.

In Guldara, a district of Kabul, more than 95% of families use wood to cook their food, warm their houses, and boil water. The efficient stoves are estimated to require 65% less fuel than traditional methods and produce significantly less smoke, helping to contribute to healthier lives for women and children who spend the most time around the stoves. By reducing the need for firewood, the stoves also help improve household work productivity by reducing the amount of time that women spend collecting firewood.

The project expects to distribute a total of 19,488 aluminum cookstoves to communities with little or no access to electricity.

↓65%

LESS FUEL REQUIRED THAN TRADITIONAL METHODS

Inhabitants
4.27 million

GDP per capita
$502

Geographic area
275 km²

THE CHALLENGE

Only a small percentage of the Afghan population have access to efficient cookstoves, and traditional mud or clay stoves are partly responsible for the 54,000 premature deaths that occur due to air pollution every year.

CO-BENEFITS

Economic

More efficient stoves will reduce cooking times by 50% and consume significantly less fuel, resulting in time and money savings.

Environmental

Reducing the need for firewood means a reduction in deforestation from forests around Kabul, conserving natural habitats and reducing carbon emissions.

Health

The project will play an important role in reducing indoor air pollution and improving the respiratory health of recipients.

Energy-efficient stoves. The locally made cookstoves are part of a larger project that aims to bring more sustainable forms of energy to Afghanistan (photo by Afghanistan Sustainable Energy for Rural Development of the Ministry of Rural Rehabilitation and Development).

Green Passport Campaign Urges Youth to Protect the Environment

Inspired by the Sustainable Development Goals (SDGs), the Ministry of Environment and Tourism of Mongolia has begun implementing the Green Passport campaign in Ulaanbaatar to increase youth participation in environmental conservation.

The "Let's Change Our Attitude" program was born from the Government of Mongolia's belief that future generations may hold the key to achieving the SDGs by creating sustainable behaviors at a young age. Incentivized by a "Green Passport," children in Ulaanbaatar were encouraged to conduct activities relating to environmental protection.

The project kicked off in 2018 when 25,000 students and 6,000 teachers were introduced to the green passport idea and were challenged to design a recycling bin for batteries. Following the competition, 300 waste battery bins with the winning design were installed, which collected 14,393 kilograms of batteries. Additional activities were planting 400 trees of over 36 different species, learning about ozone layer and ecosystems, and cleaning up 20 tons of waste.

Phase two of the campaign involved another 25,000 students living outside the capital. Among the activities was a video competition for information dissemination, a "waste marathon," environmental reporter training, and a flash-mob competition.

↓42K

TONS OF CO$_2$ EMISSIONS REDUCED BY 2030

Inhabitants
1.54 million

GDP per capita
$5,653

Geographic area
4,704 km²

THE CHALLENGE

Ulaanbaatar has experienced rapid growth in recent years, and faces challenges related to air pollution and waste management.

CO-BENEFITS

Social

The project aims to broadly change lifestyle habits among participants and has the potential to result in significant water and energy savings.

Environmental

The proper disposal of waste batteries will prevent ecological damage and pollution, and the planting of trees will increase biodiversity in the city.

Encouraging eco-friendly attitudes. The campaign will continue for another 2 years, aiming to reinforce sustainable habits while involving 30% of all Mongolian high school students (photo by Tuul Kawa).

Testing Low-Carbon Living in Jilin

A pilot community is looking to implement a variety of low-carbon concepts into city planning, construction, and the lives of residents to control rising GHG emissions.

Yuebeizhen is a suburb of Jilin that is in the midst of rapid urbanization, but is looking to explore low-carbon alternatives to conventional urban growth trajectories through the introduction of energy efficient and renewable energy infrastructure, water-saving technologies, improved solid waste management, and the promotion of low-carbon lifestyle alternatives.

A priority of the initiative is to reduce emissions resulting from fossil fuel combustion; to this end, a broad range of investments are being made. For example, residential and community buildings will be renovated to include energy-saving measures, and street lights and buildings will be powered by renewable energy. Further, the use of water-saving appliances, rainwater harvesting, and flow metering will contribute to a more secure water supply, and waste recycling programs will be implemented.

The project started in 2017 and is currently in progress, with a few measures already under development.

↓7%

REDUCTION IN CO$_2$ EMISSIONS

Inhabitants
4.12 million

GDP per capita
$4,969

Geographic area
27,120 km^2

THE CHALLENGE

Yuebeizhen is in the midst of rapid urbanization and has seen an analogous increase in GHG emissions.

CO-BENEFITS

Health

By reducing GHG emissions and various forms of waste, the project helps reduce diseases caused by pollution and improves the overall health of residents.

Economic

Investments in energy renovation, including solar street lights and building retrofits, will save the city around CNY1.35 million per year.

Social

The project will contribute to a shift in consumption and lifestyle habits, with bicycle rental stations, electric vehicle charging stations, and a low-carbon living guide being introduced.

A model for low-carbon city development. As part of its low-carbon transformation, Yuebeizhen will improve its solid waste management with recycling boxes installed in communities (photo by ADB).

"Art for Action" Inspires Younger Generation

↑1.5M

PEOPLE ENGAGED ON THE SUBJECT OF SUSTAINABILITY

Inhabitants
1.54 million

GDP per capita
$5,653

Geographic area
4,704 km²

THE CHALLENGE

Ulaanbaatar is confronted with hazardous air quality resulting from emissions and frequent water shortages, both of which will be further exacerbated by climate change.

CO-BENEFITS

Economic

Local businesses and nongovernment organizations benefited from funds raised by the exhibition and commercial merchandise based on the graphic designs.

Social

The emphasis on an environment conscious and community-centric approach inspired continued public engagement and opened the door for art to be used as a tool for advocacy.

A poster campaign in Ulaanbaatar sought to provide education about environmental issues through the medium of graphic art as part of a larger cultural movement.

The "Me+We" project utilized visual storytelling to raise awareness on the current challenges that have resulted from a period of rapid urbanization in Mongolia's urban center. Graphic art presented a simple, yet powerful tool that could be understood universally, and which could reach new audiences through the use of social and mass media outlets.

The nation's top graphic artists worked in collaboration with activists and environmental experts to create images designed to educate the population on some of the most pressing environmental issues the city faces, including air pollution and predicted water shortages.

The completed exhibit, composed of over 50 posters, launched in 2018 in the widely visited Shangri-La Centre before later traveling to the country's largest universities and neighboring regions. The way in which action today can influence the future was a focal point of the exhibition, which affirmed the important role Mongolia's younger generation has to play.

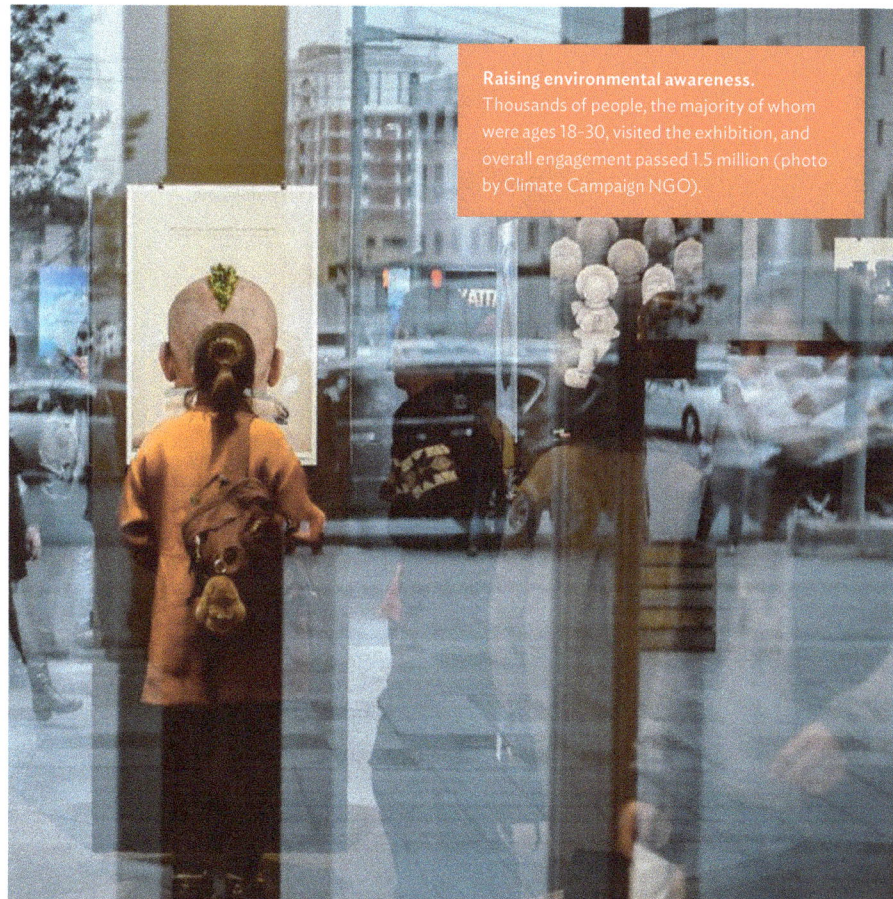

Raising environmental awareness. Thousands of people, the majority of whom were ages 18-30, visited the exhibition, and overall engagement passed 1.5 million (photo by Climate Campaign NGO).

Low-Carbon Initiatives Boost Jiangxi Communities

↓700
TONS OF CO$_2$ EMISSIONS REDUCED ANNUALLY

SHANGRAO

Inhabitants
6.83 million

GDP per capita
$5,740

Geographic area
22,791 km^2

THE CHALLENGE

Carbon emissions from Jiangxi's communities have increased in recent years due to modernization and tourism demand.

CO-BENEFITS

Environmental

Improved waste and water management have reduced pollution and enhanced the quality of the surrounding environment.

Economic

Low-carbon energy improvements will create jobs in the community and help residents save on electricity costs.

Social

Jiangxi's growing tourism sector received a boost from measures promoting low-carbon travel in the region.

A project in Jiangxi province is piloting a range of low-carbon initiatives including solar tiles, LED streetlights, biowaste processors, and more to reduce emissions and improve livability for residents.

Tangcun Village is a front-runner for contemporary developments across the energy, transportation, waste management, agriculture, and tourism sectors in Wuyuan county, in Jiangxi province's city of Shangrao. Wuyuan includes several small towns, villages, and swathes of picturesque countryside.

A range of new measures and technologies are being piloted to boost the sustainability of Tangcun Village's energy and waste management demands. To reduce emissions from energy, solar panels will be installed across rooftops, streetlights, and greenhouses. As part of this initiative, some solar tiles were also installed that have a more congruent aesthetic to match the traditional Chinese roof style. As for waste, bio-processors will produce fertilizer from kitchen and organic waste, and water-saving toilets will be installed throughout the villages.

Electric vehicle charging stations and new green spaces will be introduced to benefit residents and tourists alike. Following the experience of low-carbon community design in Tangcun Village, nine further pilots in Wuyuan have been announced.

Low-carbon growth. Tangcun is a village in the PRC province of Jiangxi where low-carbon initiatives are being piloted to explore the potential of climate action (photo by Wang Heli).

Communities Commit to Low-Carbon Transformations in the PRC

Xiangtan plans to engage in low-carbon renovations and retrofitting in 20 low-income communities to meet GHG reduction targets.

The project represents Hunan Province's first multisector approach to promoting low-carbon initiatives and will target conventional community facilities that are inefficient and utilize high levels of energy. Xiangtan's municipal government has put forward a development program, beginning in 2021, that focuses on making the city energy-efficient and resilient, and has committed to peaking carbon emissions by 2028.

Improved insulation of walls, doors, and windows will increase building efficiency, and PV and solar thermal rooftop panels will be installed. The introduction of LED street lighting, conversion from coal to natural gas cooking, and construction of electric vehicle and bicycle charging stations will further contribute to reduced emissions. These improvements and the inclusion of low-carbon features will reduce GHG emissions and improve the living environment for residents.

The pilot project is expected to spread to other cities in the PRC, and is part of ADB's $150 million loan to the Xiangtan Low-Carbon Transformation Project.

Low-carbon living for Xiangtan communities. The proposed plan takes a multisector approach and will be implemented primarily in aging communities in the Yuhu and Yuetang districts, improving the living environment for residents (photo by Nordic Sustainability).

↓9.5K

TONS OF CO_2 EMISSIONS REDUCED PER YEAR

Inhabitants 2.88 million

GDP per capita $11,371

Geographic area 5,015 km²

THE CHALLENGE

Xiangtan has experienced rapid urbanization in recent years and is the highest per capita emitter of carbon emissions in Hunan Province.[6]

CO-BENEFITS

Economic
Energy-efficient buildings will decrease utility consumption, saving money for residents and improving real estate value in the area.

Environmental
Ecosystem-based adaptation measures will enhance green spaces, increase biodiversity, and improve water quality.

Health
Reduced GHG emissions will improve local air quality, providing public health benefits and improving the quality of life of residents.

[6] G. Yu et al. 2017. Quantitative Research on Regional Ecological Compensation from the Perspective of Carbon-Neutral: The Case of Hunan Province, [PRC]. *Sustainability*. 9 (7). pp. 1–12.

Climate Adaptation on the Shores of Southeast Asia's Largest Lake

↑90%

SANITATION COVERAGE ACHIEVED

Inhabitants
Kampong Chhnang: 40,911
Pursat: 68,247

GDP per capita
Kampong Chhnang: $1,643*
Pursat: $1,643*

Geographic area
Kampong Chhnang:
5,521 km²
Pursat: 12,692 km²

* national data

THE CHALLENGE

Severe flooding and poor environmental sanitation are key infrastructure issues affecting the health and well-being of residents in Kampong Chhnang and Pursat.

CO-BENEFITS

Economic

Flood control components will decrease property damage and lessen agricultural and commercial losses in the region.

Health

Reduced flood risks and improved sanitation will decrease the prevalence of waterborne diseases, and the introduction of sidewalks and solar lamps will improve resident safety.

Improvements to drainage, flood protection, solid waste management, and community mobilization will help increase resilience for the 100,000 residents living around Tonle Sap, Southeast Asia's largest freshwater lake.

The Kampong Chhnang and Pursat provinces, which border Cambodia's Tonle Sap Lake, are introducing environmental infrastructure to improve sanitation and reduce vulnerability in the regions, which can expect to see increased flooding due to climate change.

As part of the ongoing Integrated Urban Environmental Management in the Tonle Sap Basin project, Kampong Chhnang and Pursat will increase flood protection through embankment construction and reinforcement, with sluice gate and drainage installation further strengthening efforts. These measures are expected to reduce the number of households affected by flooding in Kampong Chhnang by 80% and in Pursat by 50%.

Additionally, new solid waste collection and landfill management will improve sanitation in the two regions, and community resilience will be increased through climate change awareness campaigns and the introduction of early warning systems. Moving forward, the provision of similar decentralized urban services and resilient infrastructure may be replicated in other municipalities and urban areas within Cambodia.

The project is financed by ADB in conjunction with the Government of Cambodia.

Key infrastructure issues. Intensified flooding due to climate change will impact communities who live in low-lying areas with limited solid waste management (photo by Vuth Ratha).

Increasing Resilience in Bangladesh's Rapidly Growing Cities

↑2.5K

HECTARES OF LAND
PROTECTED FROM
FLOODING DURING
MONSOON SEASON

Inhabitants
Dhaka: 17 million
Khulna: 2 million

GDP per capita
Dhaka: $7,712
Khulna: no data

Geographic area
Dhaka: 2,161 km²
Khulna: 46 km²

THE CHALLENGE

Climate change impacts,
including heavier rainfall and
increased flooding, will stress
existing drainage and stormwater
management systems in Dhaka
and Khulna.

CO-BENEFITS

⌁ Economic

The drainage subproject will
reduce flood damage, road
maintenance costs, and earning
losses during waterlogging periods.

👥 Social

Improved roads will provide safer
and more pedestrian-friendly
traveling conditions, and will
enable residents to better access
health and education services.

♡ Health

Better sanitation practices
and enhanced solid waste
management will lower the risk of
waterborne diseases during times
of flooding.

A $223 million development project in Dhaka and Khulna will
provide climate-resilient drainage systems, urban roads, and solid
waste management to reduce flood vulnerability, increase mobility,
and improve living conditions in the region.

The Second City Region Development Project will include the rehabilitation of
300 km of urban roads around Dhaka and Khulna, two rapidly growing cities. The
inclusion of roadside drains and landscaping to absorb excess water will mitigate
some damage from flooding, allowing residents to access public services in times
of need.

Both Dhaka and Khulna are located in the vicinity of the world's largest river delta,
the Ganges Delta. Coupled with the fact that they are also close to sea level,
both are particularly vulnerable to flooding. Improvements to 150 km of drainage
systems, green slope protection, and enlarged retention reservoirs will increase
flood resilience, with the flood inundation period expected to decrease by 30%
during an average monsoon season.

A final element of the project focuses on solid waste management, with the
construction of a composting plant and biogas production facilities in Khulna.
Alongside a public awareness campaign about recycling, this approach seeks
to reduce GHG emissions, decrease water pollution, and improve sanitation
practices.

The project is assisted by a $150 million ADB loan.

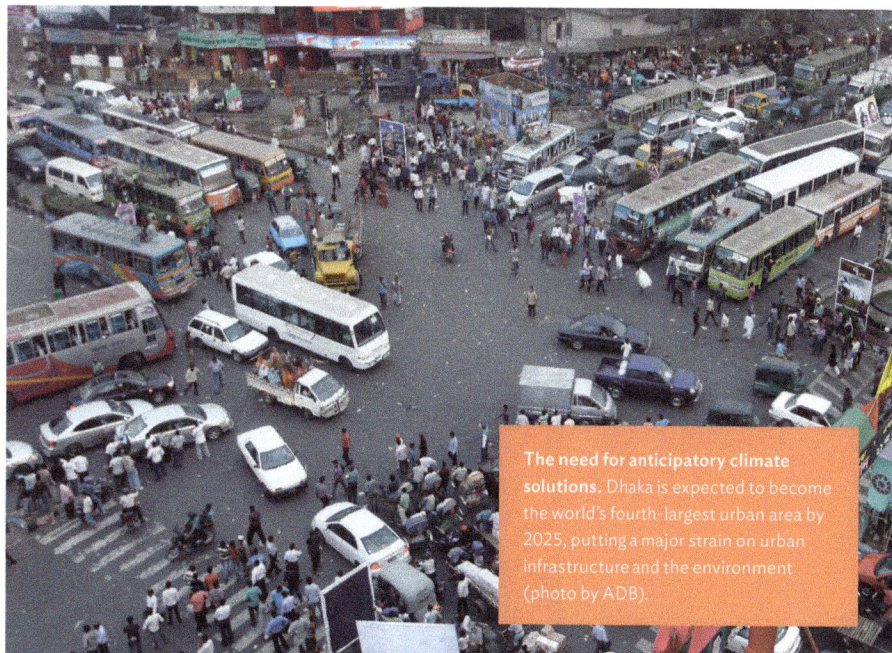

**The need for anticipatory climate
solutions.** Dhaka is expected to become
the world's fourth-largest urban area by
2025, putting a major strain on urban
infrastructure and the environment
(photo by ADB).

Post-Coal Economies for Heilongjiang

↓31K

TONS OF CO$_2$ EMISSIONS REDUCED EVERY YEAR

HEILONGJIANG

Inhabitants
37.51 million

GDP per capita
$5,236

Geographic area
473,000 km^2

THE CHALLENGE

Heilongjiang Province is rich in carbon-intensive resources, and burning them for energy has been an economic driver for people and cities living there.

CO-BENEFITS

Economic

The project will directly create an estimated 7,819 jobs, and further long-term employment opportunities in the four cities are anticipated.

Environmental

Diversification and remediation projects will reduce pollution in the area and provide residents with a more secure water supply.

The PRC's northernmost province has long been a furnace for the country's coal-powered economy. Now in transition, the cities of Hegang, Jixi, Qitaihe, and Shuangyashan are searching for more sustainable alternatives.

As the PRC begins to transition toward more sustainable fuels, a just transition must unfold in northern states with coal-powered economies to avoid unemployment and economic stagnation. The $1 billion, 5-year program has two main components: financial stimuli for small and medium-sized enterprises (SMEs) and investment in environmental restoration for coal-related impacts.

The first component focuses on boosting local businesses with business development services for SMEs in the four cities. The support services will help improve business planning, research and development, and financial management, meaning local businesses can offer jobs for those no longer working in coal.

The environmental restoration component will target coal mines, such as the open-pit mine in Hegang, which will be remediated through a series of soil treatments to make it suitable for vegetation and the construction of green spaces. In Qitaihe and Shuangyashan, waste-rock dumpsites will be remediated and reused as open spaces.

The project is funded by $310 million in loans from ADB and $220 million from the European Investment Bank.

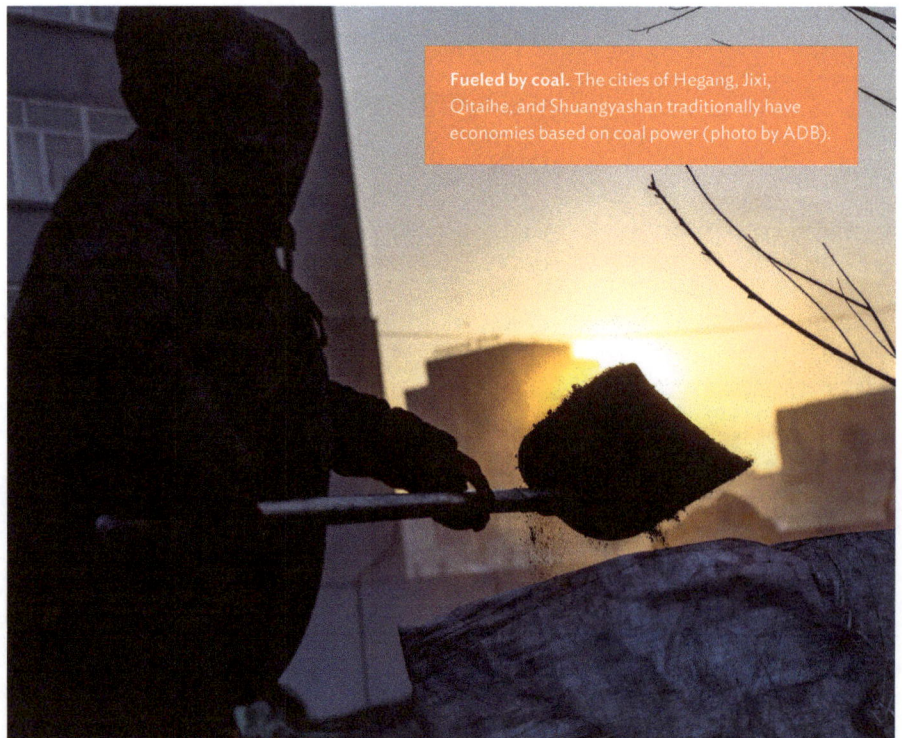

Fueled by coal. The cities of Hegang, Jixi, Qitaihe, and Shuangyashan traditionally have economies based on coal power (photo by ADB).

Putting Livability at the Center of Adaptation and Mitigation

↓60K

TONS OF CO$_2$ EMISSIONS REDUCED EVERY YEAR

Inhabitants
556,400

GDP per capita
$8,329

Geographic area
1,748 km²

THE CHALLENGE

Yanji suffers from inefficient public transport, traffic congestion, unsafe conditions for pedestrians, lack of green spaces, and vulnerability to flooding.

CO-BENEFITS

Economic

The project will create employment opportunities, and will result in increased savings in vehicle operating costs, travel times, and wastewater treatment costs.

Social

Improved urban mobility, road safety, and city greening will create a more secure and pleasant living environment for Yanji residents and tourists.

Environmental

Sponge city infrastructure and expanded green spaces will improve surface water quality, reduce air and soil pollution, and provide carbon sequestration for the city.

Efforts to improve livability and address climate change in Yanji have a two-pronged focus on improving public transportation and resilience with "sponge city" designs.

Yanji borders the Democratic People's Republic of Korea in an area that is particularly vulnerable to flooding. This is a significant problem given that this region will likely see increases in precipitation above even the high national average.[7]

This 7-year, $250 million initiative plans to integrate low-carbon and climate-resilient urban development measures to reduce emissions and provide flood protection in Yanji. It will feature the first bus rapid transit (BRT) system in the area and an emphasis on green landscaping, which will connect key areas of the city and decrease carbon emissions, while also addressing traffic and public safety issues.

Landscaped paths, bicycle lanes, and river greenways will provide more attractive routes to bus stops and will increase the city's capacity to retain and filter stormwater. This sponge city infrastructure will be integrated with an improved drainage pipe network based on a hydraulic model that simulates future storm scenarios, resulting in reduced urban flooding risks.

The project is supported by an ADB loan of €117.5 million (about $130 million).

---

[7] Intergovernmental Panel on Climate Change. 2018. The People's Republic of China Third National Communication on Climate Change.

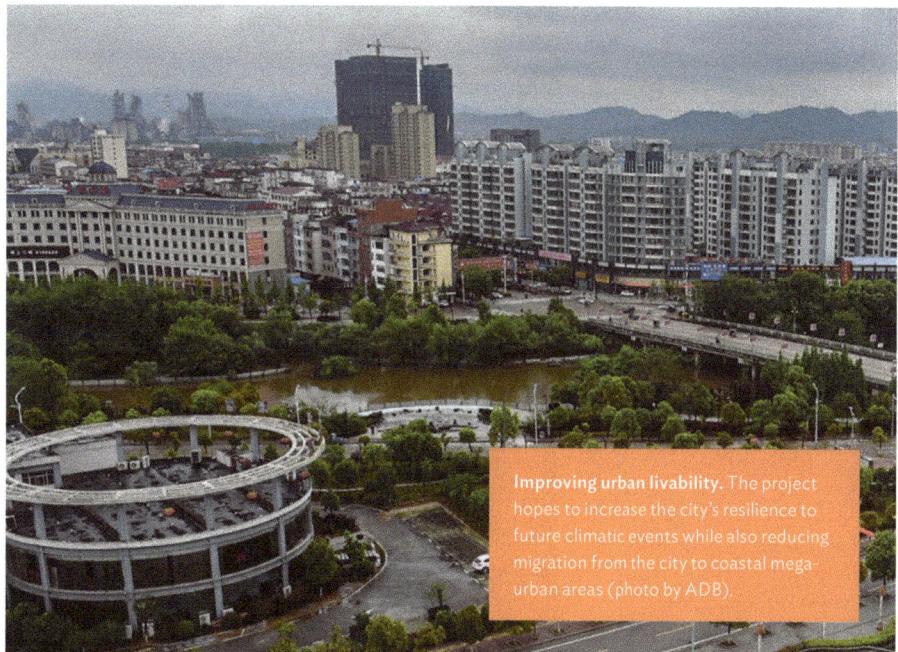

Improving urban livability. The project hopes to increase the city's resilience to future climatic events while also reducing migration from the city to coastal mega-urban areas (photo by ADB).

Tamil Nadu Invests in Climate-Resilient Infrastructure

A huge infrastructure investment program, covering 10 cities in the state of Tamil Nadu, aims to implement climate-resilient water supply, sewerage, and drainage systems across the region.

India's southern state of Tamil Nadu wants to promote climate-resilient urban development, as the lack of piped water and sewerage networks serves to increase the vulnerability of residents to water-related climate change impacts. Efforts will center around improving access to water, sanitation, and wastewater disposal services across 10 cities, while also reducing pollution and CO_2 emissions.

India's first solar-powered sewage treatment plant will be installed in Coimbatore on a pilot basis, and connected with more than 2,800 km of new sewage collection pipelines. Wastewater reuse for industrial purposes will also be made possible by advanced treatment facilities and the flood resilience of the systems will be increased through improved drainage and the raising of critical infrastructure.

Smart water management systems will also be installed, with 1,500 km of water supply distribution systems. Along with 40 new water storage reservoirs, this will provide residents with a reliable source of water, and the introduction of 110 district metered areas will reduce nonrevenue water loss in water-scarce regions.

The program is funded through an ADB loan of $500 million, an Asian Clean Energy Fund grant of $2 million, a government counterpart financing of $766.4 million, and an ADB technical assistance grant of $1 million.

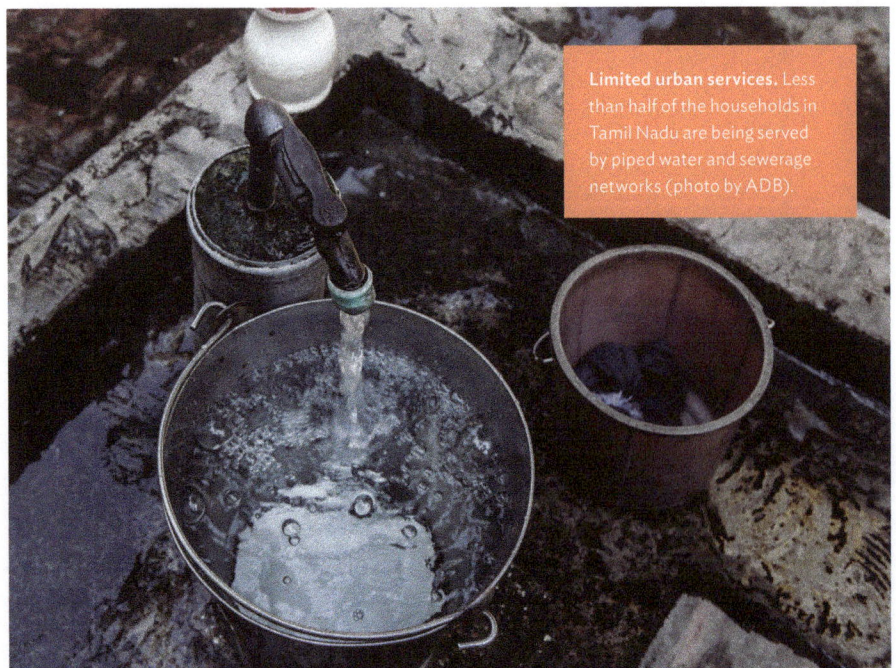

↓3.4K

TONS OF CO_2 EMISSIONS REDUCED EVERY YEAR

TAMIL NADU

Inhabitants
72.15 million

GDP per capita
$2,862

Geographic area
130,058 km²

THE CHALLENGE

Rapid urbanization has been accompanied by infrastructure deficits, polluted waterways, and increased vulnerability to droughts and urban floods.

CO-BENEFITS

♡ Health

The absence of sewerage systems in project areas poses a major health risk, which improved infrastructure will help mitigate.

Social

The program will directly address water, sanitation, and drainage service gaps among disadvantaged and underserved communities, including in slums.

Environmental

Improved sanitation and wastewater management will reduce contamination and ensure that residents have access to a clean water supply.

Limited urban services. Less than half of the households in Tamil Nadu are being served by piped water and sewerage networks (photo by ADB).

Climate-resilient infrastructure. Water infrastructure is a key component of climate adaptation. Cities in the Indian state of Tamil Nadu are investing in water infrastructure that is resilient to climate change impacts and can also help the city become less vulnerable to natural hazards (photo by ADB).

Overcoming Water Scarcity in the South Pacific

↑95%

OF RESIDENTS IN THE GREATER HONIARA AREA SERVED BY PIPED WATER SUPPLY BY 2047

Inhabitants
84,520

GDP per capita
$2,197

Geographic area
22 km²

THE CHALLENGE

Urban population growth in the Greater Honiara Area, including five urban towns of Auki, Gizo, Noro, Munda, and Tulagi, has resulted in decreasing rates of access to basic urban services such as water supply, sanitation, solid waste collection, and drainage.

CO-BENEFITS

Health

Health outcomes for the population, especially those living in informal communities, will be improved through better access to water and sanitation services.

Social

The program will include gender design features that benefit women and girls, such as menstrual hygiene awareness and quotas in the project workforce and water user groups.

Environmental

The introduction of a reed bed wastewater treatment plant and increased efficiency of the water supply system will result in GHG reductions.

Solomon Islands is preparing for a future characterized by intense rainfall events and extreme droughts, both of which will strain the country's insufficient and aging water and sanitation systems.

The scarcity of freshwater sources and lack of sanitation infrastructure are two significant problems for residents of Solomon Islands, the vulnerability of whom will be further exacerbated by climate change impacts, with both increased rainfall variation and extreme droughts expected by 2050.

The urban water and sanitation sector project will feature upgraded water treatment facilities, two new reservoirs, rehabilitation of 10 km of water supply pipes with added leak detection, and new metered connections. These measures will reduce water loss by more than 30%, while also increasing supply capacity and security.

Improvements to sanitation services will also be incorporated, including new pump stations, a reed bed wastewater treatment plant, repairs to aging pipes, and the expansion of the existing network to include 3,000 new households. As a whole, this initiative will contribute to water conservation efforts aimed at protecting communities against an increasingly uncertain future.

The $92 million project is funded in part by a concessional loan and grant from ADB, with cofinancing from the European Union and World Bank.

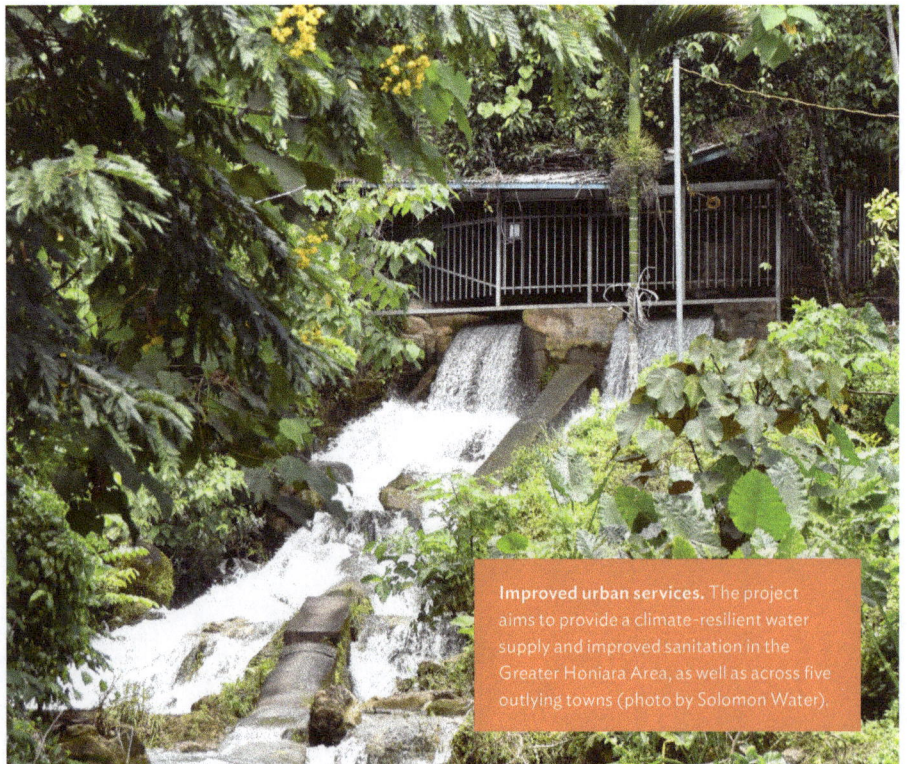

Improved urban services. The project aims to provide a climate-resilient water supply and improved sanitation in the Greater Honiara Area, as well as across five outlying towns (photo by Solomon Water).

A Fresh Approach to Water for a Future with Rising Tides

↑40K

ADDITIONAL CUBIC METERS OF FRESHWATER SUPPLY PER DAY

Inhabitants
315,947

National GDP per capita
$6,220

Geographic area
97 km²

THE CHALLENGE

Rising tides and increasing droughts will increase pressure on Suva's water supply, which has not kept pace with recent levels of migration to the city.

CO-BENEFITS

♡ Health

Improvements to water and sewage pipelines will limit water contamination and groundwater seepage, as well as health risks from waterborne diseases.

Economic

Reduced water losses and increased supply capacity will lower costs for users and help accommodate population growth in the area.

Environmental

Water savings, sewer systems, and the capture of methane from an anaerobic digester will reduce emissions by 26,000 tCO_2e per year.

The capital city of the small island nation of Fiji is improving the delivery of freshwater and the processing of used water, while also acting on implications of climate change.

The islands of Fiji are on the frontline of climate change and expect to see rising sea levels over the coming decades unless strong mitigation measures are implemented internationally and quickly. An investment program will seek to provide clean drinking water for the inhabitants of Suva and improve the management of wastewater in the city.

A new raw water intake on the Rewa River, water treatment plant, pumping station, reservoir, and transmission main will increase freshwater capacity in Suva and connect to the existing system. As a climate adaptation response to predicted sea level rise and increased risk of saltwater intrusion, the intake has been moved 20 km further upstream. A catchment management plan will be developed to safeguard the water source and ensure long-term sustainability.

Suva's wastewater network will also be rehabilitated and expanded by upgrading 31 existing wastewater pumping stations, upgrading existing infrastructure, and expanding the network to service an additional 15% of households.

The $405 million investment program is funded in part by an ADB loan.

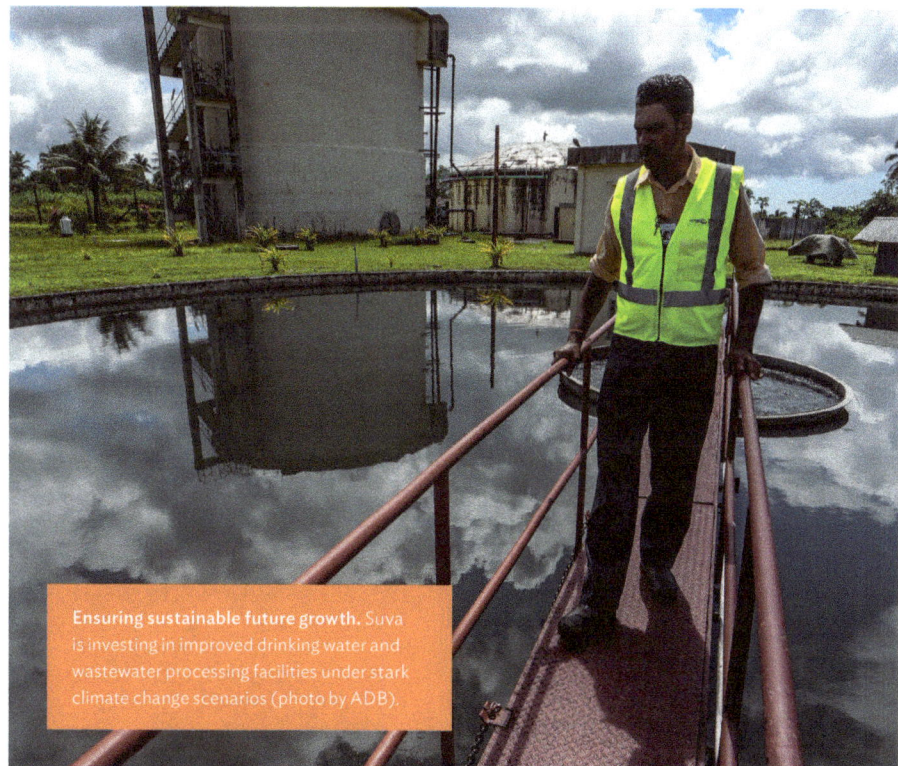

Ensuring sustainable future growth. Suva is investing in improved drinking water and wastewater processing facilities under stark climate change scenarios (photo by ADB).

Solving Kiribati's Water Puzzle

Kiribati is installing the largest desalination plants in the Pacific Islands as part of a larger package of measures supported by ADB aimed at tackling the impacts of climate change on freshwater supplies in South Tarawa.

South Tarawa, the capital city, is installing two new desalination plants with energy offset by solar power as part of a larger program to improve access to drinking water and reduce waterborne diseases, especially in times of drought or during disasters.

There are just two freshwater sources in South Tarawa that have a combined output of around 2,000 cubic meters per day. Combined with leakages from the existing water supply network, it is estimated that just 11 liters per person per day are available to the population, well below the 50-liter level that is recommended to meet minimum health requirements.

The 7-year program will start by rehabilitating the existing water supply infrastructure to reduce water leakages, followed by the installation of two desalination plants that will have their energy consumption offset by a 2.5 MW PV solar system.

Other outputs within the program will focus on building the capacity of the Ministry of Infrastructure and Sustainable Energy and Public Utilities Board to better manage water supply infrastructure.

The project cost is estimated at $62 million, with financing provided in part by ADB, the World Bank, and the ADB-administered Green Climate Fund.

↑6K

CUBIC METERS OF FRESHWATER OUTPUT PER DAY

Inhabitants
56,338

National GDP per capita
$1,655

Geographic area
203 km²

THE CHALLENGE

Overcrowding and inadequate water, sanitation, and hygiene coverage are major issues in highly urbanized South Tarawa. Further exacerbating the situation, its main freshwater sources are threatened by climate-induced inundation and prolonged droughts.

CO-BENEFITS

Health

Access to a safe and climate-resilient water supply will eliminate the high dependence on contaminated water and thus reduce waterborne diseases.

Economic

The plant's use of solar energy will result in cost savings, and households that no longer need to boil water prior to consumption will save time and fuel.

Social

An awareness-building campaign about water, sanitation, and hygiene practices will include gender-sensitive communication and outreach.

Safe, reliable, and climate-resilient water supply. The new desalination plants will be the largest constructed in the Pacific Islands (photo by ADB)

Dushanbe Doubles Down on Water Security

Tajikistan's capital is seeking to incorporate climate-resilient measures in its water supply and sanitation infrastructure to increase the water security of its 863,000 residents.

Tajikistan, located among the mountains of Central Asia, is one of the region's most climate-vulnerable countries facing rainfall variation, extreme droughts, and disappearing glaciers. In Dushanbe, around 60% of water is currently lost due to leaks or theft, operation is intermittent at only 4–8 hours a day, and contamination is frequent.

The initiative aims to provide safe drinking water and improved sanitation in urban areas, a vital task in a city facing increasing water shortages under climate change. To do so, district metering areas will be introduced to reduce water loss through active leak detection and smart meters. Existing wells, pumps, and 17 km of transmission main will also be rehabilitated, and protection will be put in place to safeguard groundwater well fields.

Flood risk management and drainage will also be enhanced, with an increased storage capacity of 4,500 cubic meters. Further, improvements to a sewage collector in the southern part of the city will reduce contamination risks, especially during flood periods.

The $45 million project is made possible by a grant provided by ADB.

↑100K

RESIDENTS HAVE IMPROVED ACCESS TO A SAFE WATER SUPPLY

Inhabitants
863,400

GDP per capita
$877

Geographic area
203 km²

THE CHALLENGE

Rapid population growth and expansion have increasingly strained the city's aging and poorly maintained water supply and sanitation infrastructure.

CO-BENEFITS

♡ Health

Improvements to sanitation services will reduce the incidence of waterborne illnesses like ascariasis and diarrhea in the project areas.

⌁ Economic

Targeted consumers will receive cost savings from non-incremental water consumption with the switch to piped water from alternative sources.

Mounting urban pressure. Dushanbe has seen increased migration, contributing to environmental degradation, poor livability, and limited economic prospects (photo by State Executive Authority of Dushanbe City).

Dhaka Prepares for Reduced Freshwater Availability

↓10%

NONREVENUE WATER

Inhabitants
17 million

GDP per capita
$7,712

Geographic area
2,161 km²

THE CHALLENGE

Access to clean water is a challenge for residents, who are often forced to rely on unreliable and expensive illegal water lines and private vendors.

CO-BENEFITS

Health

The provision of a safe and reliable water supply will reduce water-related health risks, especially among children.

Economic

Piped water and metered connections will reduce the cost of water, with consumers no longer needing to pay high rates or water purification costs.

By improving the provision and reliability of Dhaka's water supply, the city also hopes to become more climate-resilient.

Providing sufficient drinking water is a challenge for the city of Dhaka, with 17 million residents and still growing by 3.6% annually. Climate change is expected to exacerbate threats to the city's vulnerable water supply, with altered precipitation patterns, higher temperatures, sea level rise and salinization, river contamination, and increased pollution from flooding, all contributing to a decline in availability. These threats combined have necessitated the city to take urgent action.

Under the initiative, district metered areas will be expanded, more than 1,500 km of water distribution networks will be rehabilitated, and new technologies such as automated meter reading, supervisory control and data acquisition (SCADA), and a water quality monitoring system, will be installed, effectively reducing nonrevenue water loss.

The construction of 9,500 new and legalized water points will also provide water at a lower cost for low-income communities and help Bangladesh's largest city adapt to water uncertainties in a future under climate change.

The project is assisted by a $275 million loan from ADB.

Water supply network improvement. The project will build on previous efforts to improve piped water supply and reduce physical water losses in Dhaka (photo by Md. Arifur Rahman).

Bhutan Battles Rising Waters in the Himalayan Foothills

↑15K

PEOPLE HAVE ACCESS TO RECLAIMED LAND THAT IS SAFE AND SECURE

Inhabitants
27,658

GDP per capita
$3,438

Geographic area
15.6 km²

THE CHALLENGE

Located between the Himalayan foothills and Amochhu River, safe development space is at a premium in Phuentsholing.

CO-BENEFITS

Economic

The development provides new space for public services and commercial businesses, allowing the city to diversify its economy and provide employment opportunities for residents.

Social

The township will be developed in an inclusive manner, with land allocated for social housing and infrastructure that is designed with gender-sensitive measures in mind.

Four kilometers of flood protection river wall has opened up 66 hectares of new, safe land for Bhutan's second-biggest city, located in a high-risk area of the Himalayan mountains.

The Phuentsholing Township Development Project seeks to protect the city from increasing floods, riverbank erosion, and landslides, while also allowing for safe urban expansion to take place. New housing is often built precariously on steep hillsides and floodplains around the city, making residents vulnerable to natural hazards.

To address these challenges, the city is planning a number of developments to increase resiliency against floods and provide new areas for low-risk housing. Following a hydrological evaluation of the Amochhu river, the city was able to design a 4 km flood protection wall designed to protect against 100-year flood events. By installing the wall, 66 hectares of previously high-risk floodplain has been reclaimed for urban development.

As extra precautions for the new township in Phuentsholing, the city also raised the ground level, implemented extra drainage infrastructure, and is rolling out an early warning system for the whole city in case of extreme floods.

Funding for the $63 million project is provided in part by ADB in the form of a concessional loan and grant.

Bhutan's busiest border city. Phuentsholing is located on the border with India in the Himalayan foothills of southern Bhutan (photo by Sonam Phuntsho).

Climate Action Plans
and Inventories

→ The climate action sector presents a range of cross-sector plans, initiatives, and long-term projects that cities are using to lower their carbon footprint, inform urban planning, and improve citizen health. Impacts of climate change will be felt for generations to come, so it is essential that cities adopt action plans that align current action with future goals, including the nationally determined contributions established under the 2015 Paris Agreement.

Wuhan's climate action plan. Wuhan is aiming to peak its emissions in 2022, 8 years ahead of the PRC's national target. This type of action plan can help cities decouple emissions from economic growth in good time (photo by Tian Miao).

100 More Days of Clean Air in Chengdu

↓11.4M

TONS OF CO_2 EMISSIONS
REDUCED FROM 2013 TO 2017

Inhabitants
16.58 million

GDP per capita
$15,012

Geographic area
14,335 km²

THE CHALLENGE

Chengdu's rapid industrialization and urbanization has accompanied decades of rising GHG emissions and deteriorating air quality.

CO-BENEFITS

Economic

Reductions in air pollution and GHG emissions were not matched by economic stagnation, and the city's GDP grew by 8% each year.

Health

The policies enacted under this action plan have significantly improved air quality in Chengdu, increasing the life expectancy of residents by 3.9 years.

The number of "good air quality" days in Chengdu has increased by 100 and life expectancy also rose by an estimated 3.9 years after 5 years of air pollution action in this megacity in the PRC.

To address emissions and poor air quality in Chengdu, the city spent over CNY1.5 billion over 5 years on a range of actions targeting high-polluting sectors.

Starting in 2013, the 5-year plan took a five-step approach: problem identification, solution research, design, implementation, and monitoring. Although the city's population grew by 12% over the 2013–2017 period, the policies implemented led to significant reductions in air pollution and GHG emissions, including a 54% reduction in sulfur dioxide (SO_2) and 35% reduction in particulate matter ($PM_{2.5}$).

Much of the improvement in Chengdu's air quality is due to a sharp drop in the city's reliance on coal, with a combination of hydropower, solar, wind, and natural gas being used instead. A number of policies have also been introduced to help shift toward cleaner forms of mobility, as well as control dust emissions from the expanding urban construction sector. There is also ongoing research into nature-based solutions along highways to reduce pollutants from vehicles.

Chengdu inspires climate action. Knowledge sharing was a crucial element in this project, and this plan to improve air quality has since been replicated in other cities in the region (photo by Liu Wei).

DUSHANBE, TAJIKISTAN

Launching Emission Monitoring in Tajikistan

↓24%

REDUCTION IN CO_2 EMISSIONS EXPECTED BY 2030

Inhabitants
863,400

GDP per capita
$877

Geographic area
203 km²

THE CHALLENGE

With growing levels of private vehicle usage in Dushanbe, air pollution is a primary concern for the government.

CO-BENEFITS

Environmental

Data from the project will be used to develop a new traffic management scheme with recommendations to reduce the environmental impact of vehicles.

Health

Efforts to reduce emissions from private vehicle use will result in a long-term reduction of air pollution in the city.

As road transportation in Tajikistan grows in popularity, carbon emissions are predicted to rise. The country's capital Dushanbe has begun the very first monitoring scheme as a first step toward reducing emissions from the transport sector.

The city of Dushanbe has begun to measure road pollution with a monitoring scheme as a first step to gain control over its emissions and improve livability for its citizens.

With just 863,400 residents, the relatively small capital city is taking efforts to better understand traffic flow, intensity, and emissions levels from the city. This information allows cities to determine where they should direct mitigation efforts, and the data is vital in creating policies and new infrastructure that can improve air quality, safety for pedestrians, and lower emissions from the transportation sector.

Tajikistan is the lowest contributor of GHG emissions in Central Asia in both absolute and per capita emissions, and is responsible for just 0.02% of global emissions. However, levels of private vehicle usage in Dushanbe have grown, and in 2014 it was estimated that the transportation was responsible for 13 times as much air pollution as the industrial and energy sectors. The city plans to adopt targeted measures aimed at investing in alternative fuels and improved vehicle efficiency.

Data-driven decision-making. Monitoring emissions from cars produces data that can be used to make informed decisions about air pollution and environmental protection strategies (photo by Sayfullo Qoridov).

Batumi's Efficiency Action Plan

The city of Batumi is in the midst of an energy efficiency push aimed at reducing local emissions by 20%.

Following a multi stage energy action plan focusing on efficiency improvements, Georgia's third-largest city has reduced GHGs by over one-fifth in key emitting sectors throughout the city.

Batumi established a Sustainable Energy Action Plan, which involved constructing an inventory of emissions from the key sectors, defining measures for emissions reductions, and monitoring those reductions. These were compared against a business-as-usual scenario which allowed the city to estimate the amount of emissions avoided through the project.

The plan also includes local capacity building and awareness raising, aiming to enhance the knowledge of citizens about the rise of local CO_2 emissions.

Energy-saving bulbs, a methane extraction system for the town's landfill, and solar panels for municipal buildings are some of the actions taken across the various sectors. Overall, the city estimates that the implemented activities have cut emissions by 22% compared to a business-as-usual scenario.

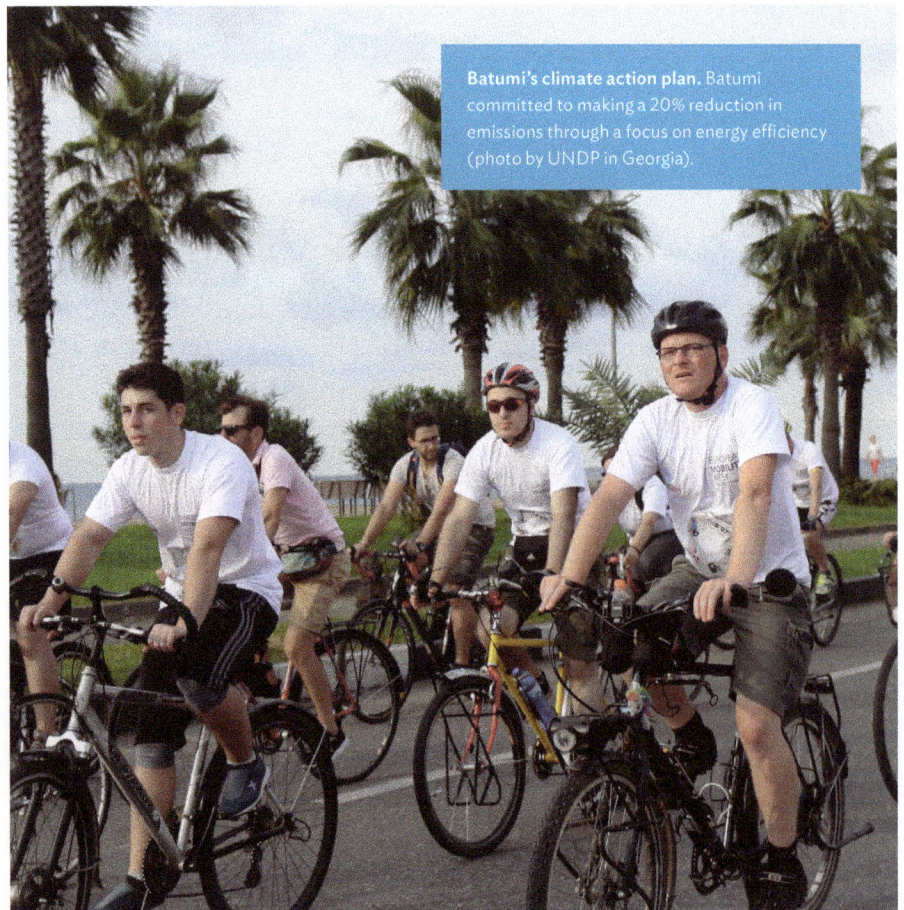

↓22%

REDUCTION IN CO_2 EMISSIONS

Inhabitants
169,100

GDP per capita
$3,852

Geographic area
200 km²

THE CHALLENGE

High levels of economic growth and increase in tourism have not been matched by sustainable urban development measures in Batumi.

CO-BENEFITS

Economic

The introduction of more efficiency and energy-saving measures will result in savings for both commercial businesses and individuals.

Environmental

Improvements to the transport sector, added thermal insulation, and climate control systems will all contribute to the reduction of emissions.

Batumi's climate action plan. Batumi committed to making a 20% reduction in emissions through a focus on energy efficiency (photo by UNDP in Georgia).

Peaking Emissions 8 Years Ahead of Schedule

The Wuhan Model looks across sectors to peak emissions in the megacity by 2022, 8 years ahead of the national target.

The so-called Wuhan Model targets the six areas of industry, energy, life, ecology, infrastructure, and demonstration for low-carbon developments and initiatives. Some highlights from the far-reaching model include low-carbon agriculture, priority to non-fossil energy, low-carbon transportation, greening of the city, and creating low-carbon demonstration projects to improve the reach of the project and scale it to other cities. It is the first carbon emissions peak action plan proposed at the city level in the PRC.

In 2018, the city was on course for emissions of 155 million tCO_2e, but following the initiatives across the different sectors, the city estimates to have reduced over 16 million tCO_2e. This has also benefited the city's air quality, which has seen a 35% improvement compared with 2013.

As Wuhan and the PRC's economies continue to grow, it is vital that they can decouple emissions from growth to help mitigate climate change. By 2022, the city aims to peak at 173 million tons, after which, it aims to decrease them toward 2030 and beyond.

↑41%

GREEN COVERAGE RATE IN URBAN AREAS

Inhabitants
11.21 million

GDP per capita
$20,960

Geographic area
8,569 km²

THE CHALLENGE

As a large city, Wuhan has previously experienced an increase in the growth rate of urban carbon emissions.

CO-BENEFITS

Social

The low-carbon transformation of the city will improve livability with more efficient public transport, green spaces, and improved infrastructure.

Environmental

The project will reduce coal consumption and emissions in Wuhan, with corresponding pollutants falling by 19% compared to 2015.

Health

Reduced pollution will result in lesser deaths related to poor air quality by about 50,000.

Wuhan Carbon Emissions Peak Action Plan. The action plan is composed of six major projects and can be used as an example of successful low-carbon development for other cities in the PRC (photo by Fan Jianjun).

Xiangtan's Public Procurement Approach

The public sector can exert a huge influence on private activity. By adopting a green procurement approach, Xiangtan is using its influence to create positive change.

As part of Xiangtan's larger green transformation program, the city has introduced a green low-carbon procurement policy. Xiangtan is the first city in the PRC to introduce a green procurement policy with proper institutional setup for execution and implementation. By 2022, the aim is to incorporate the green procurement policy into its fully functioning e-procurement system, with integrated low-carbon procurement data and monitoring and reporting functions.

Green public procurement is a process where public authorities search for goods, services, and works that have a reduced environmental impact throughout their life cycle. It is a voluntary policy tool that many other cities throughout the world have successfully implemented to cut GHG emissions that the city is responsible for.

In 2018, Xiangtan was selected by the National Development and Reform Council as a low-carbon city, and this is one of several programs being implemented in a drive to peak local emissions by 2028.

The introduction of the green low-carbon procurement policy is part of Xiangtan's low-carbon policy and institutional reforms supported by the ADB's policy-based loan of $50 million.

↓240K

TONS OF CO₂ EMISSIONS REDUCED OVER 10 YEARS

Inhabitants
2.88 million

GDP per capita
$11,371

Geographic area
5,015 km²

THE CHALLENGE

Xiangtan has experienced rapid urbanization and growth over the past decade, which corresponds to increased consumption and emissions.

CO-BENEFITS

Economic

The project promotes local green and low-carbon technologies and suppliers, and consequently will contribute to local economic growth.

Environmental

The range of measures presented in this plan contribute to overall emission reductions for the city, which will reduce air, water, and soil pollution.

Social

By creating lifestyle and consumption habits that emphasize resource conservation, Xiangtan will be able to sustainably grow its economy and society.

Catalyzing green procurement. It is hoped that Xiangtan can serve as a model of green procurement for other public institutes, businesses, and cities across the PRC (photo by Haiping Yu).

Ulaanbaatar Makes Concrete Plans to Address Environmental Challenges

↑14

NEW POLICY ACTION AREAS TO IMPROVE THE ENVIRONMENT

Inhabitants
1.54 million

GDP per capita
$5,653

Geographic area
4,704 km²

THE CHALLENGE

Citizens in Ulaanbaatar are faced with high levels of air pollution, insufficient urban infrastructure, and poor water quality.

CO-BENEFITS

Economic

Actions related to affordable green housing, utility development, and a circular economy will create a myriad of economic and employment opportunities.

Social

The promotion of rainwater harvesting will allow households in unplanned *ger* settlements to collect drinking water, thus becoming more self-sustainable.

Environmental

Nature-based solutions and green infrastructure provide important ecosystem services in the form of increased biodiversity and soil permeability.

The city's Green Action Plan lays out 14 tangible project ideas and implementation steps aimed at addressing priority environmental pressures and improving the well-being of all citizens.

Mongolia's capital city currently finds itself in a somewhat paradoxical situation in terms of climate change vulnerabilities, as Ulaanbaatar is located in a water-scarce area but also regularly faces severe flooding. The city also suffers from poor air quality, primarily due to the use of coal-based heat and inefficient energy systems.

Ulaanbaatar has chosen to prioritize the areas of energy, building efficiency, and land use to decrease pollution, reduce the carbon footprint from heating, and increase the quality of life for residents. To do so, the city has put forth a multisector development plan with a proposed cost of $712 million. It features a compendium of actions that address the unique environmental challenges faced by the city, with concrete targets and estimated costs included.

The 14 action areas include provisions for green housing and ecodistricts, development of a mass transit system, nature-based solutions and "sponge city" design, and improved waste management and sanitation systems.

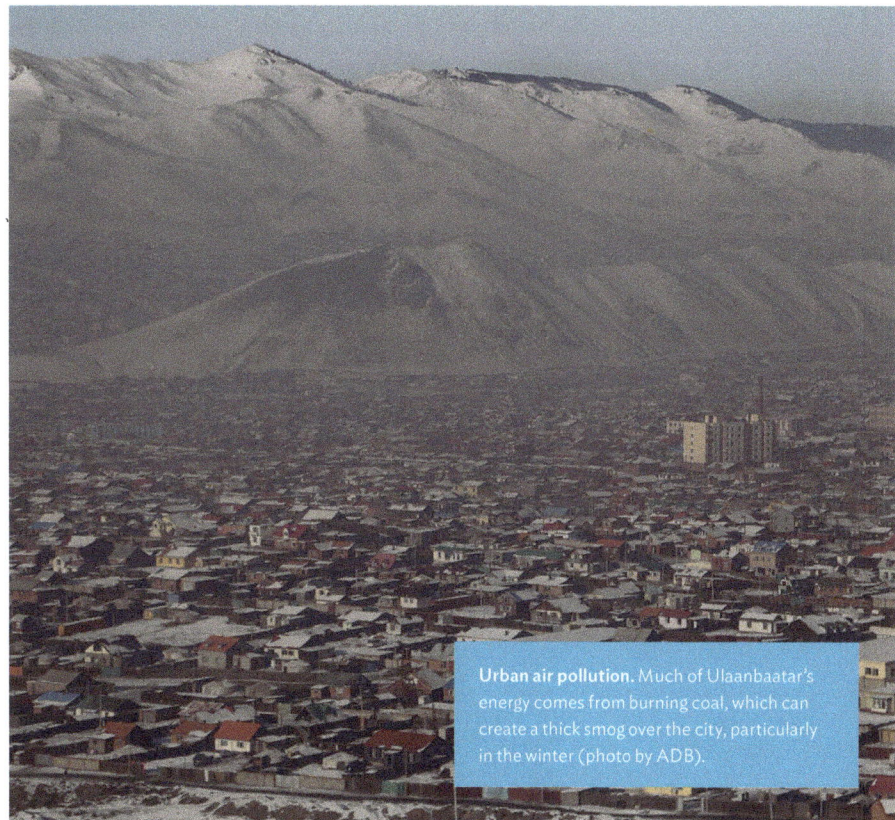

Urban air pollution. Much of Ulaanbaatar's energy comes from burning coal, which can create a thick smog over the city, particularly in the winter (photo by ADB).

Choosing from the Climate Resilience Toolbox

Xiangtan is using a newly developed toolbox to enhance collaboration for climate change adaptation planning. In particular, the city is using the toolbox to become more flood-resilient.

The Xiangtan Climate Resilient City Toolbox is an interactive platform designed for a more collaborative approach to adaptation design. It can be used at design workshops by policy makers, planners, designers, and practitioners for informed decision making of priority areas for climate resilience, defining requirements, setting adaptation targets, selecting specific interventions, and assessing effectiveness. More importantly, the use of the toolbox created a new culture of collaborative city planning within the Xiangtan Municipal Government, replacing the practice of working in silos and utilizing single sector-driven development and planning.

A climate risk and vulnerability assessment was conducted in Xiangtan to create a customized toolbox for city planners. The toolbox contains information on the cost and effectiveness of 43 ecosystem-based adaptation solutions, which can be easily selected for a project area using the toolbox's user-friendly interface.

The toolbox was used to help improve the flood resilience of Xiangtan's First Traditional Chinese Medicine Hospital, by identifying ecosystem-based adaptation solutions that could provide 7,840 cubic meters of storage capacity to protect against flooding.

The project is funded as part of ADB's technical assistance to prepare the Xiangtan Low-Carbon Transformation Sector Development Program loan project.

Web-based platform for urban resilience planning. The Climate Resilient City Toolbox has been used for a number of projects including designing the floodproofing for the first Traditional Chinese Medicine Hospital (photo by E-waters Environmental Science and Technology (Shanghai) Co., Ltd.).

↑**7.8K**

CUBIC METERS OF WATER STORAGE CAPACITY IDENTIFIED

Inhabitants
2.88 million

GDP per capita
$11,371

Geographic area
5,015 km²

THE CHALLENGE

There are many existing guidelines for climate adaptation strategies, but this can be a maze for urban planners tasked with responding to location-specific risks.

CO-BENEFITS

Environmental
Ecosystem-based adaptation measures included in the toolbox integrate the use of biodiversity and ecosystem services.

Economic
Proper implementation of policies and projects tested with the toolbox is expected to increase real estate values and avoid economic loss induced by future flood events.

Health
Accidents and the loss of lives due to flood events will be minimized in project areas.

Reducing Emissions, Improving Air Quality

A policy-based loan program was implemented to simultaneously tackle Ulaanbaatar's air pollution and reduce emissions through consistent policy objectives, regulatory frameworks, new technologies, and economic incentives.

In 2018, levels of harmful fine particulate matter in Mongolia's capital were 40 times the daily limit recommended by the World Health Organization. Rapid urbanization has resulted in the growth of vast peri-urban *ger* areas in the city that lack adequate public services and rely on the combustion of coal for heating and cooking, contributing to severe air pollution and GHG emissions.

Ulaanbaatar chose to adopt an air quality improvement plan that encompasses three multisector reform areas: a stronger regulatory framework, including preferential taxation for cleaner fuels; a program for financing air quality improvement; and stricter energy efficiency standards. These were accompanied by a public awareness campaign linking fuels, air pollution, and health.

Tangible measures related to pollution reduction and health protection also included the banning of raw coal and temporary replacement with cleaner-burning semi-coke briquettes, longer-term options for decentralized and renewable heating, and improved air filtration in schools and hospitals to reduce indoor air pollution.

The policy-based loan was sourced from ADB's ordinary capital resources.

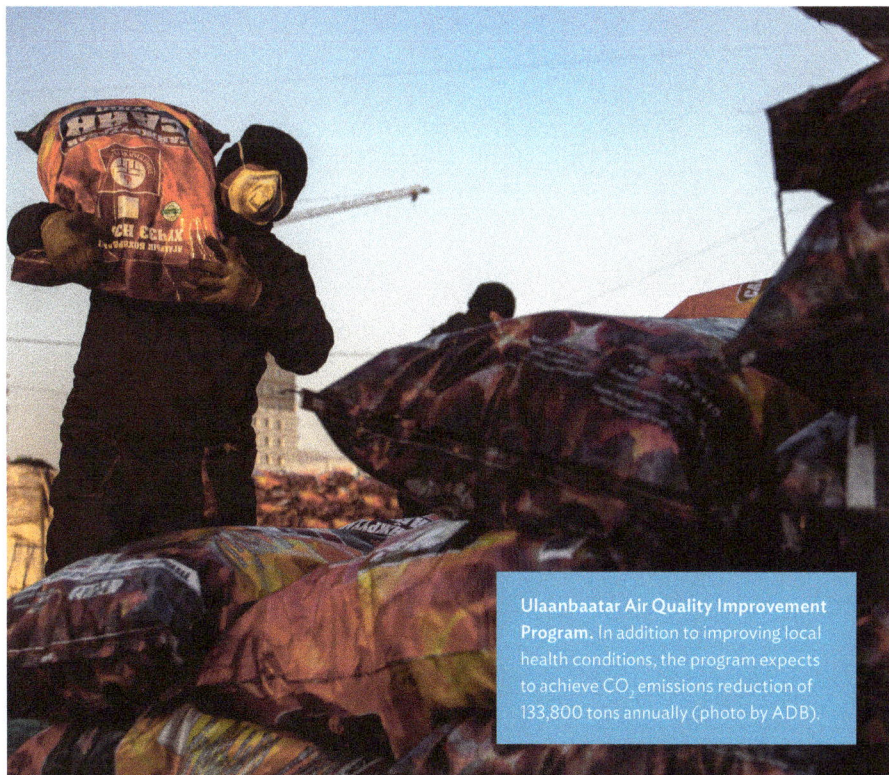

↑**20**

CONCRETE POLICY ACTIONS TO ADDRESS THE POLLUTION PROBLEM AND CUT EMISSIONS

Inhabitants
1.54 million

GDP per capita
$5,653

Geographic area
4,704 km²

THE CHALLENGE

Ulaanbaatar is one of the most polluted cities in the world and has experienced rapid urban development in recent years.

CO-BENEFITS

♡ Health

By switching to cleaner energy sources, air quality will be improved, reducing the prevalence of cardiovascular and respiratory diseases.

Social

Coordinated urban and energy planning will allow for the sustainable development of *ger* areas.

Economic

Income loss from restricted economic activities due to poor air quality will be reduced.

Ulaanbaatar Air Quality Improvement Program. In addition to improving local health conditions, the program expects to achieve CO_2 emissions reduction of 133,800 tons annually (photo by ADB).

Building Energy Efficiency

→ Retrofitting aging buildings, incentivizing efficiency upgrades, and building to the highest environmental standards are all ways that cities in Asia and the Pacific are changing the built environment to drive down emissions. Through these cost-efficient measures, cities can also cut energy bills and reduce local air pollution levels.

International Innovation Inspires Building Standards in Viet Nam

The Deutsches Haus in Ho Chi Minh City is a flagship green building in Viet Nam. The 25-floor, high-rise building combines German design and technology with local culture to cut emissions and promote international collaboration.

Opened in 2017, Deutsches Haus, or "the German House," is a 25-story building complex in Ho Chi Minh City that follows some of the highest environmental standards available, showcasing how technology and architectural design can create more sustainable high-rise buildings in urban metropoles.

Several construction choices were instrumental in earning the Leadership in Energy and Environmental Design (LEED) Platinum certification, one of the most prestigious building certifications available and the first building in Viet Nam to receive it.

Features include LED lights, hospital-grade air filtration systems, localized temperature control, and rooftop solar panels. Rainwater harvesting is in place, together with a greywater flushing system, and wastewater for cooling.

A double-layer exterior made from glass enhances natural air flow and insulation, helping keep the building interior cool in the midst of the hot and humid climate, while reducing the need for CO_2-intensive air conditioning.

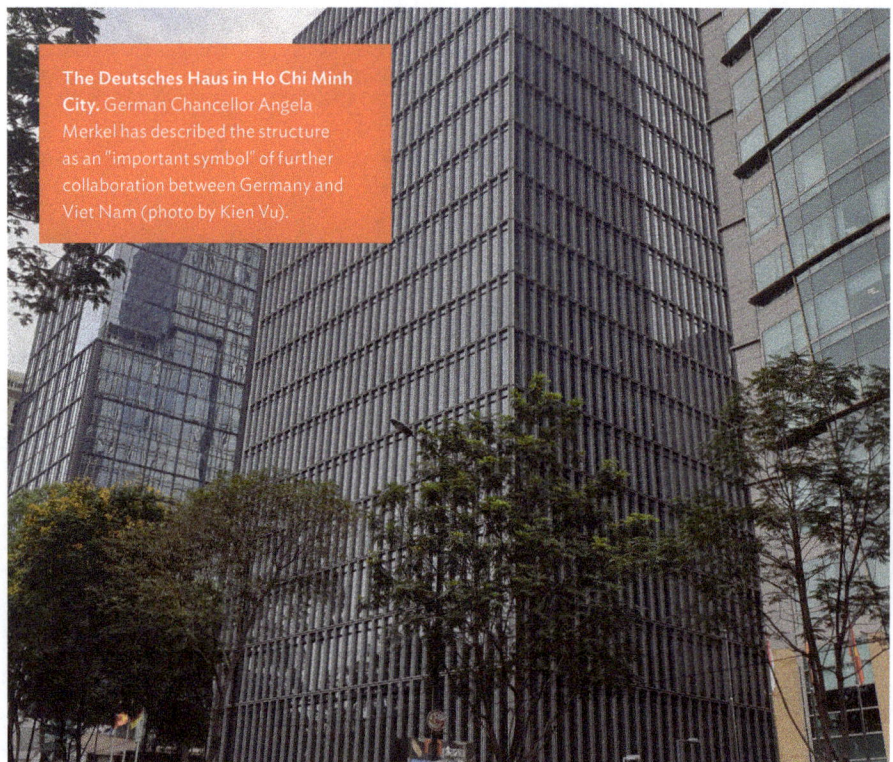

↑**25**

STORIES CERTIFIED WITH LEED PLATINUM

Inhabitants
8.99 million

GDP per capita
$6,862

Geographic area
2,061 km²

THE CHALLENGE

Ho Chi Minh City's rapid urban growth has been a driver of increased energy consumption and GHG emissions. Buildings play an important role, with an estimated 30% of final energy consumption attributed to buildings in 2015.

CO-BENEFITS

Economic

The building utilizes energy and water-saving strategies, meaning lower utility consumption and bills.

Environmental

The Deutsches Haus reduces water and energy consumption through design features that meet high building certification standards.

The Deutsches Haus in Ho Chi Minh City. German Chancellor Angela Merkel has described the structure as an "important symbol" of further collaboration between Germany and Viet Nam (photo by Kien Vu).

Traditional Medicine Meets Low-Carbon Design

Xiangtan's new traditional medicine hospital will integrate green building design, clean energy technologies, and ecosystem-based adaptation measures to make the structure low-carbon and climate-resilient.

Xiangtan's new traditional medicine hospital has applied passive building design by incorporating better building materials and water- and energy-saving features, resulting in 20% savings in water and energy compared to other conventional hospitals. The new traditional medicine hospital will be the first certified hospital in the PRC for Excellence in Design for Greater Efficiencies (EDGE)—a green building certification. In addition, the enhanced use of ecosystem-based adaptation measures on the hospital premises will reduce the vulnerability of the hospital to floods and droughts, both of which are expected to increase in the region due to climate change.

To optimize the energy efficiency of the planned hospital, a natural gas-based combined cooling, heating, and power generation (CCHP) system will be utilized. The CCHP system will be integrated with a rooftop solar system to significantly improve the energy efficiency of the hospital. Moreover, the CCHP can be operated off-grid, which helps make the hospital resilient to power outages . Planners have embraced an ecosystem-based adaptation design to provide flood protection and increase the overall resilience of the structure. A range of rainwater retention measures will be employed, including rainwater gardens and harvesting, permeable pavement, green roofs, and retention ponds, with a total water storage capacity of 7,840 cubic meters.

ADB provided a loan to cover the $98 million cost of the Xiangtan new traditional medicine hospital, which is a part of ADB's $150 million loan to the Xiangtan Low-Carbon Transformation Project.

↓3.2K

TONS OF CO_2 EMISSIONS REDUCED ANNUALLY

Inhabitants
2.88 million

GDP per capita
$11,371

Geographic area
5,015 km²

THE CHALLENGE

Xiangtan's recent urban growth has been accompanied by an increase in GHG emissions, and the city's new hospital will be built in a flat and flood-prone area.

CO-BENEFITS

Economic

The EDGE-certified measures will save the hospital 4,400 MWh of energy and 27,800 cubic meters of water each year, resulting in cost savings of $580,000.

Health

Green zones around the hospital will improve local air quality and provide recreational space for staff, patients, and visitors.

Environmental

Ecosystem-based adaptation measures will provide more green spaces, which will strengthen biodiversity and serve as a source of medicinal plants and herbs.

Xiangtan's new traditional medicine hospital. The building will be the PRC's first hospital to receive Excellence in Design for Greater Efficiencies (EDGE) certification (photo by Fang Yang).

Low-Carbon Housing for 35,000 Residents of Ulaanbaatar

↓**200K**

TONS OF CO₂ EMISSIONS
REDUCED ANNUALLY

Inhabitants
1.54 million

GDP per capita
$5,653

Geographic area
4,704 km²

THE CHALLENGE

Residents in Ulaanbaatar living in *ger* areas often lack access to critical public infrastructure and rely on heavily-polluting and inefficient coal-based heating.

CO-BENEFITS

♡ Health

Improved wastewater collection will reduce health risks from pollution and the project will contribute to improving air quality, reducing currently high rates of respiratory illnesses.

Social

Up to 35,000 people will obtain affordable housing and public spaces with adequate provision of clean water, waste management facilities, and heating.

Environmental

The project will reduce air pollution and increase energy efficiency through a shift in heating sources, and improve environmental conditions by reducing pit latrines and wastewater discharge, creating more livable urban areas.

A housing project in Mongolia's capital aims to provide low-carbon homes for residents of the city's peri-urban areas and hopes to inspire future development.

This large-scale demonstration initiative will focus on the development of 10,000 affordable housing units equipped with rooftop solar panels, better insulation, and improved connectivity to the central energy grid, water supply, and sanitation services. The housing will be contained within a new, resilient eco district.

The Green Affordable Housing Project, running until 2027, focuses on improving the lives of inhabitants who currently live in Ulaanbaatar's *ger* and peri-urban areas. The traditional yurt settlements are vulnerable to climate change and are hot spots for GHG emissions and air pollution in the city due to the use of inefficient coal stoves.

Insulation and the use of a central grid supply will reduce demand for heating, while also increasing energy efficiency. PV solar panels will supply zero-emission electricity, and pollution will be further reduced with the installation of modern toilet facilities. The new housing will also feature smart monitoring systems, grey water recycling, and rainwater harvesting, and long-term debt financing will be made available to incentivize participation in the low-carbon housing market and provision of affordable green mortgages.

The project has a budget of $570 million, and is partly financed by $175 million in loans and $53 million in grants from ADB and the Green Climate Fund.

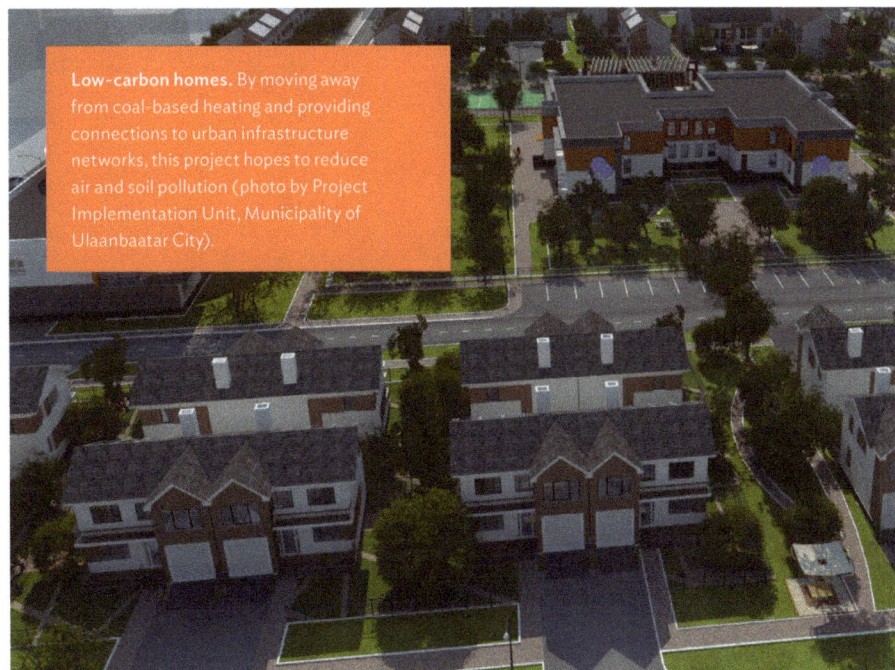

Low-carbon homes. By moving away from coal-based heating and providing connections to urban infrastructure networks, this project hopes to reduce air and soil pollution (photo by Project Implementation Unit, Municipality of Ulaanbaatar City).

ADB Headquarters Goes for Gold

ADB has implemented several retrofits to its headquarters in Manila to reduce energy and water consumption, lower GHG emissions, and win LEED Gold certification.

Over the past 6 years, the ADB headquarters has undergone a number of changes to improve the environmental performance of the building in Mandaluyong City, Metro Manila. The changes have resulted in a recertification under the United States Green Building Council's LEED Existing Building Operations and Maintenance—achieving the LEED Gold certification.

In 2014, ADB switched the energy that powers the building away from conventional sources to renewables. Now, the building's power comes from a mix of geothermal (96%) and solar energy (4%) from PV panels on the roof.

There has also been a focus on water, with internal recycling and rainwater harvesting helping to reduce demand for potable water, as well as water efficient appliances reducing water consumption.

As a result of the building changes made, annual emissions totaled 12,349 tCO_2e in 2019, 39.84% lower than in 2013, the baseline year for GHG reporting. Since 2016, ADB has offset the remaining emissions by purchasing carbon credits from ADB-supported renewable energy projects.

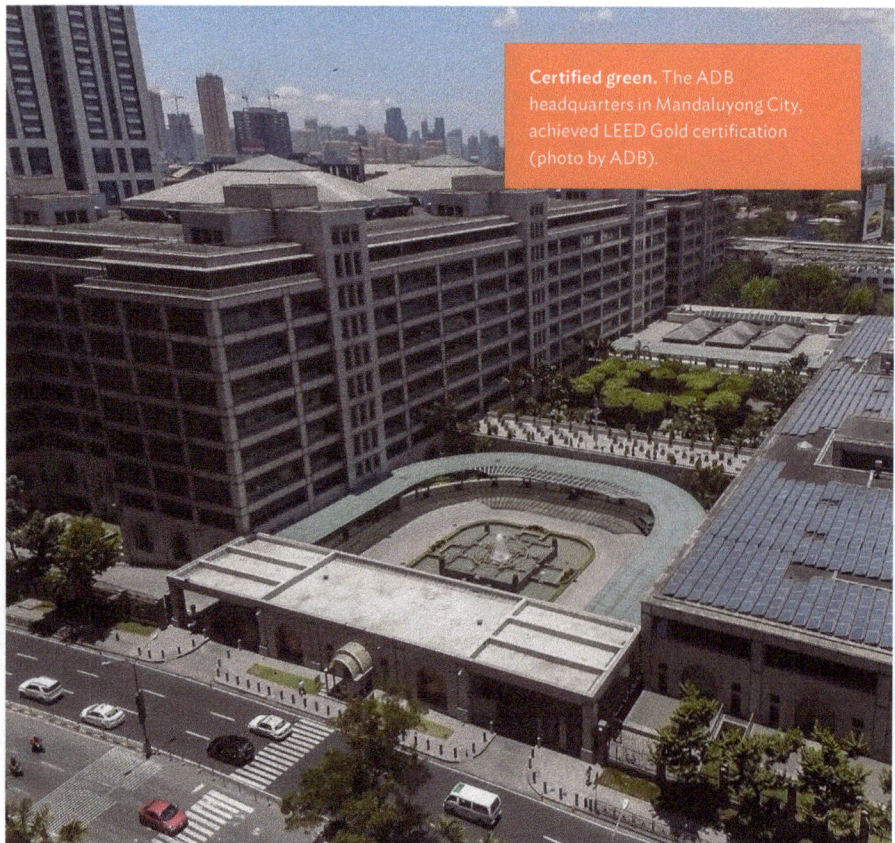

↓47%

REDUCTION IN CO_2 EMISSIONS FROM 2013 TO 2016

Inhabitants
1.78 million

GDP per capita
$7,100

Geographic area
39 km²

THE CHALLENGE

As a highly urbanized area, Mandaluyong experiences environmental problems often associated with urbanization such as air and water pollution, flooding, solid waste management issues and climate change.

CO-BENEFITS

Economic

Reduced power intensity as well as lower water and paper consumption has resulted in reduced operational expenses.

Social

ADB has committed to employ best practices, safe operating procedures, and appropriate technologies to eliminate health and safety risks to its personnel at the workplace.

Certified green. The ADB headquarters in Mandaluyong City, achieved LEED Gold certification (photo by ADB).

Retrofits Keep Heat In and Emissions Down

↓7K

TONS OF CO$_2$ EMISSIONS REDUCED ANNUALLY

Inhabitants
1.54 million

GDP per capita
$5,653

Geographic area
4,704 km²

THE CHALLENGE

The energy sector is responsible for around 70% of Mongolia's GHG emissions, of which the building sector accounts for 11% through heat and electricity consumption. This highlights the need for energy efficiency improvements in both new and existing buildings throughout the city.

CO-BENEFITS

Social

Private sector participation is expected to create over 1,000 new jobs in both construction and manufacturing sectors in Ulaanbaatar.

Health

The project will reduce air pollution through decreased energy demand in the buildings that have coal-based heating, improving air quality for citizens.

Ulaanbaatar is targeting leaky building facades with a retrofitting program to improve insulation and decrease the heating required in the cold Mongolian winters.

The Mongolian capital is targeting insulation improvements in some of the central apartment buildings that are over 30 years old and extremely inefficient.

By targeting 375 buildings in the first phase with improved wall and ceiling insulation, as well as installing triple glazed windows, the city estimates they can reduce almost 7,000 tCO$_2$e emissions annually. The savings will primarily come from reduced demand for coal-intensive heating and power.

For the first phase, the government is focusing on retrofitting 375 building blocks over the next 5 years at a cost of $72 million. In preparation for the start of the scaling up of the building retrofitting project, the municipal government has begun a demonstration project retrofitting with an initial 51 building blocks, at a cost of $2.5 million.

If the project goes well, the city could look to implement similar retrofits to the rest of the 1,072 building blocks.

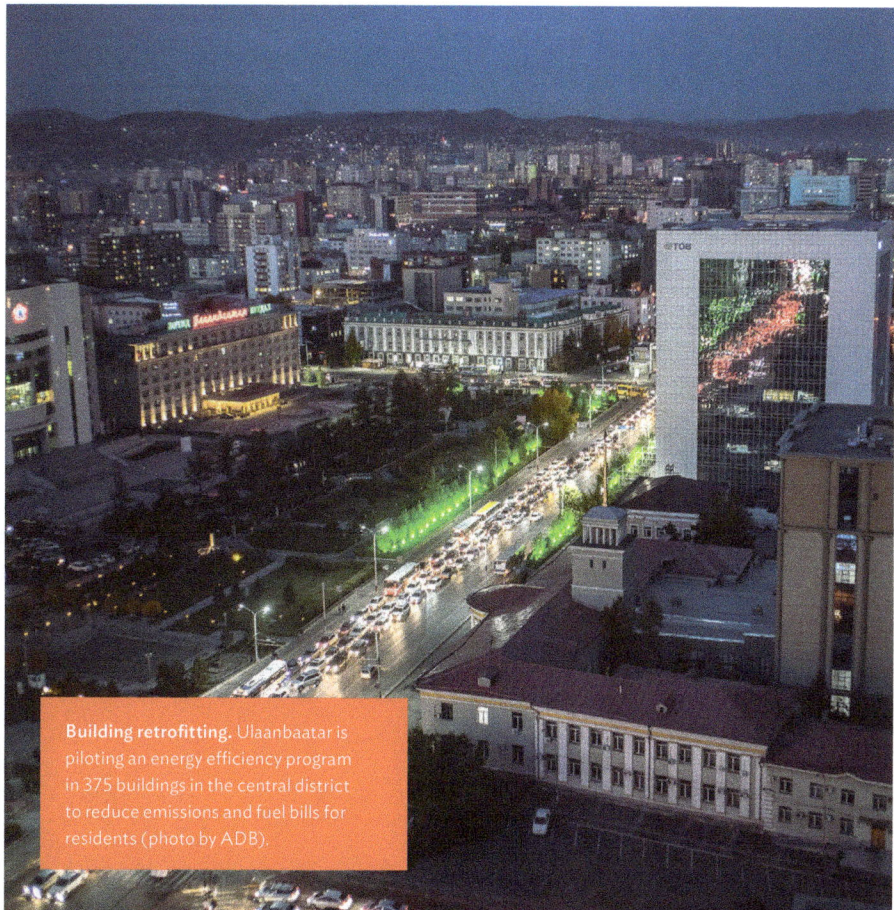

Building retrofitting. Ulaanbaatar is piloting an energy efficiency program in 375 buildings in the central district to reduce emissions and fuel bills for residents (photo by ADB).

A Helping Hand for Local Kyrgyz Efficiency Projects

↑$2M

IN LOANS AVAILABLE TO BUSINESSES

Inhabitants
1.05 million

GDP per capita
$1,309

Geographic area
170 km²

THE CHALLENGE

The Kyrgyz Republic saw massive construction of residential and public buildings in the 1960s and 1970s, with a high concentration built in Bishkek. These buildings are inefficient and carbon-intensive, and still make up the majority of buildings in the city today.

CO-BENEFITS

Economic

By investing in projects with low environmental risks, the program stimulates economic activity among the population.

Social

The program has provided loans to 734 households to make efficiency improvements, which will also raise the quality of life for residents.

A $35 million fund backed by the European Bank for Reconstruction and Development (EBRD) is helping the Kyrgyz Republic home and business owners to take energy and water efficiency improvements into their own hands, cutting bills and emissions.

Home and business owners throughout the Kyrgyz Republic can benefit from generous loans, grants, and technical assistance to improve energy and water efficiency, cutting bills as well as carbon emissions from the small and mountainous nation. The Kyrgyz Sustainable Energy Financing Facility (or KyrSEFF+) initiative is a $35 million fund backed by the EBRD that supports projects ranging from wall insulation to solar energy and rainwater harvesting in homes in the Kyrgyz Republic.

To make sustainable investments attractive for the people of the Kyrgyz Republic, KyrSEFF+ loans are supported by grant incentives of up to 35% and can be paired with technical assistance, where local partner banks make services easily accessible.

Businesses can also access funding: up to $2 million, with up to 15% in grants. These can help boost business through financing of efficient production and management systems, such as machinery for agribusinesses and wastewater treatment systems for hotels.

So far, 734 households and 74 businesses have been supported by loans and grants.

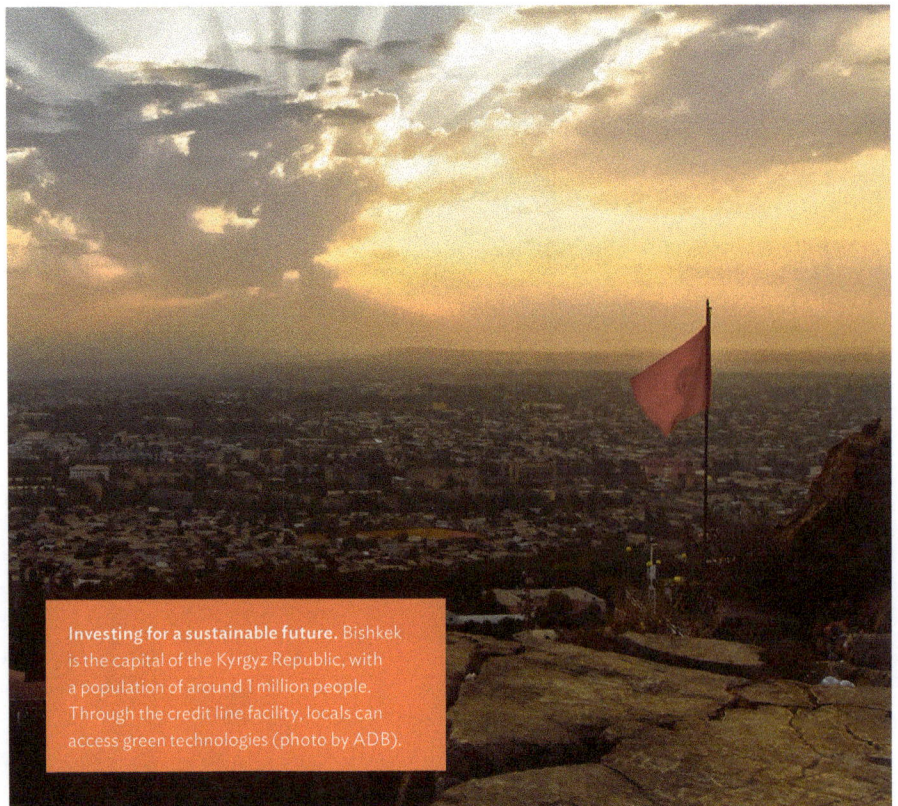

Investing for a sustainable future. Bishkek is the capital of the Kyrgyz Republic, with a population of around 1 million people. Through the credit line facility, locals can access green technologies (photo by ADB).

Eco Village in Mongolia Replaces Coal with Sunshine for Heat

↓50%

REDUCTION IN CO_2 EMISSIONS

Inhabitants
1.54 million

GDP per capita
$5,653

Geographic area
4,704 km²

THE CHALLENGE

Residents living in informal settlements rely on coal-based heating facilities, which are one of the main contributors to harmful air pollutants in the city.

CO-BENEFITS

♡ Health

As part of a range of initiatives being undertaken to improve air quality in Ulaanbaatar, the project will contribute to decreasing rates of pneumonia and other respiratory illnesses.

Social

Participating households will have access to a reliable energy supply and improved urban services like waste management.

A project in the Khoroo 19 subdistrict of Ulaanbaatar aims to deliver affordable and green housing solutions that reduce the reliance on coal for heating.

The Tsaiz Eco Village will provide 176 households with solar thermal heating systems and efficient insulation. The aim is to reduce energy demand in the winter and offer a more sustainable alternative to burning coal for warmth, which will also help to improve air quality in the city.

The majority of Ulaanbaatar's population lives in *ger* districts, which refer to Mongolia's traditional circular felt housing units, and require heating for 8 months out of the year, with temperatures falling below –30°C. The Tsaiz Eco Village will make use of the abundant sunshine that Mongolia receives year-round to provide decentralized and low-carbon heating solutions for the semi-detached properties in the form of solar thermal panels.

Further improvements to waste management will ensure that soil, water and air pollution in the area are reduced, improving livability for residents. The construction of the eco village is expected to be completed in 2021, and if successful, similar technologies may be deployed in other *ger* districts.

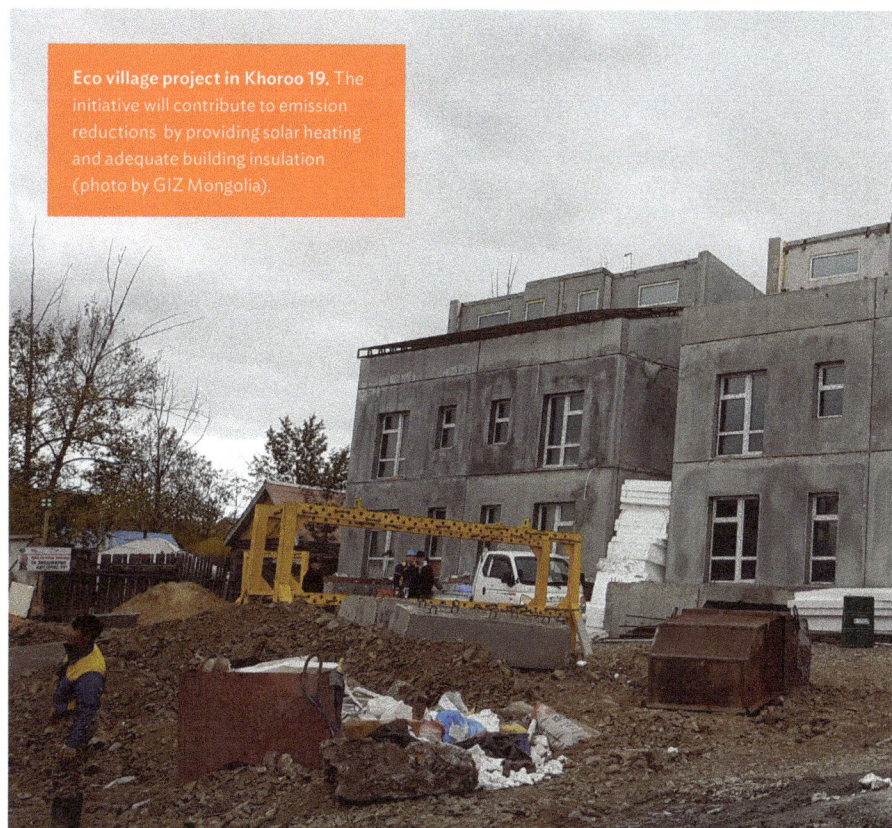

Eco village project in Khoroo 19. The initiative will contribute to emission reductions by providing solar heating and adequate building insulation (photo by GIZ Mongolia).

Solid Waste

→ In many cities in Asia and the Pacific, rapid levels of urbanization have created a gap between sustainable waste treatment needs and capacity. To move away from a reliance on polluting landfills and illegal dumping, cities are taking steps to manage waste more responsibly and adopt the three "Rs": reduce, reuse, and recycle.

The need to focus on urban waste. Waste is a vital area of infrastructure that cities such as Malé in Maldives must focus on to reduce their climate impact (photo by Ariel Deveza Javellana).

Confronting Khujand's Landfill Challenges

A waste treatment plant will be constructed in the industrial area of Khujand to relieve the city's problematic dependence on landfilling, creating energy for the town in the process.

The EU-funded construction of a waste treatment plant in Tajikistan's second-largest city will soon be underway. Through thermal processing, the facility will dispose of urban waste and contents from surrounding landfill sites, which have been persistent sources of environmental pollution and health hazards, and have undermined efficient land management.

In addition to establishing a plant that can process up to 45,000 cubic meters of sewage and 100 tons of solid waste, the project will also generate energy for the town and enable the reclamation of 80 hectares of land.

The project responds to an urgent need for expanding and modernizing waste treatment infrastructure in Khujand. The gap between waste treatment needs and capacities has encouraged both legal and illegal landfilling over recent decades, producing some 70 official landfills across the country, which cover 296 hectares and store 12 million tons of waste. The new plant will hopefully displace landfilling practices and encourage project iterations in other Tajik cities.

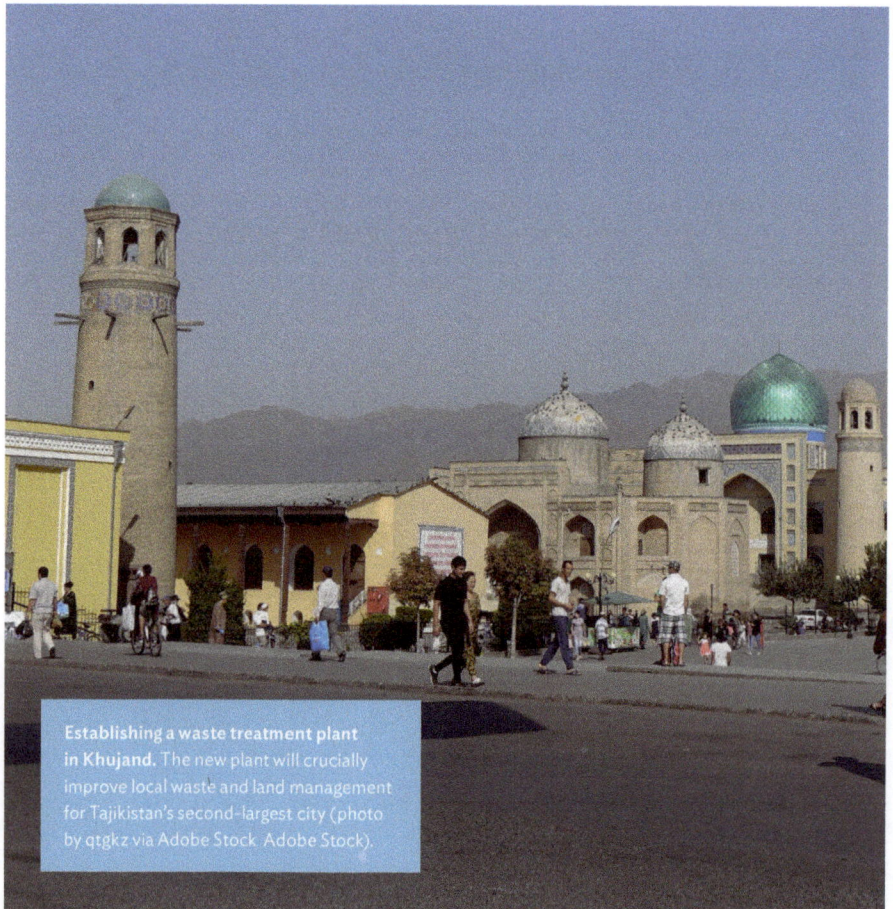

↑**80**

HECTARES OF LAND RECLAIMED

Inhabitants
181,600

GDP per capita
$3,354

Geographic area
40 km²

THE CHALLENGE

Dating back to the Soviet Union era, the antiquated and under-funded waste management infrastructure has been unable to cope with current waste flows.

CO-BENEFITS

Economic

The construction of the new facility and associated processes will create new jobs for inhabitants in Khujand.

Environmental

By improving waste processing and generating energy, the project will reduce emissions and improve the environmental conditions of the downstream Syr Darya River area.

Establishing a waste treatment plant in Khujand. The new plant will crucially improve local waste and land management for Tajikistan's second-largest city (photo by qtgkz via Adobe Stock. Adobe Stock).

Turning the Tide on Plastics

In a move to promote a circular economy and tackle marine plastic pollution, leading global food and beverage brands are ramping up sustainability efforts to increase recycled plastic content in their packaging.

The project will support Indorama Ventures Public Company Limited (IVL), a global integrated polyethylene terephthalate (PET) manufacturer and recycler, for the expansion and upgrading of its plastic recycling plants in India, Indonesia, the Philippines, and Thailand.

These recycling facilities will treat post-consumer PET bottles and convert them into feedstock for new PET bottles or downcycle them into polyester fibers. With better incentives for collection of used PET bottles, waste collectors can help reduce plastics released into the environment.

The project's innovative recycling technologies and process transformations that include energy efficiency, water conservation, waste heat recovery, and integration of renewable energy technologies such as rooftop solar will result to high-impact, low-carbon recycling plants. Targeted to be fully operational by 2022, these upgraded recycling facilities will ensure that a large number of these bottles are diverted from waste annually.

The $300 million project is funded through various financiers—$50 million from ADB, $50 million from the ADB-administered Leading Asia's Private Infrastructure Fund, $150 million from the International Finance Corporation, and $50 million from Deutsche Investitions- und Entwicklungsgesellschaft. The blue loan will follow the Blue Natural Capital Financing Facility Blue Bond Guidelines and Green Loan Principles.

↓40.4K
TONS OF CO_2 EMISSIONS REDUCED ANNUALLY

Inhabitants*
Tangerang
(Indonesia): 2.27 million
General Trias
(Philippines): 314,303
Nakhon Pathom
(Thailand): 920,030
Rayong
(Thailand): 64,256

GDP per capita
Tangerang: $5,560
General Trias: $3,270**
Nakhon Pathom: $9,080***
Rayong: $29,068***

Geographic area
Tangerang: 164 km²
General Trias: 81 km²
Nakhon Pathom: 2,168 km²
Rayong: 17 km²

*location in India to be determined
**regional data
***provincial data

THE CHALLENGE

Non-degradable, single-use plastics coupled with inefficient solid waste management and inadequate recycling systems have led to growing marine plastic pollution with Asia accounting for more than 80% of the total release of plastic into the ocean.

CO-BENEFITS

Economic

The increased recycling capacity will result to improved urban infrastructure and employment generation.

Environmental

The project will contribute to reduced GHG emissions and improved ocean health with less plastic pollution.

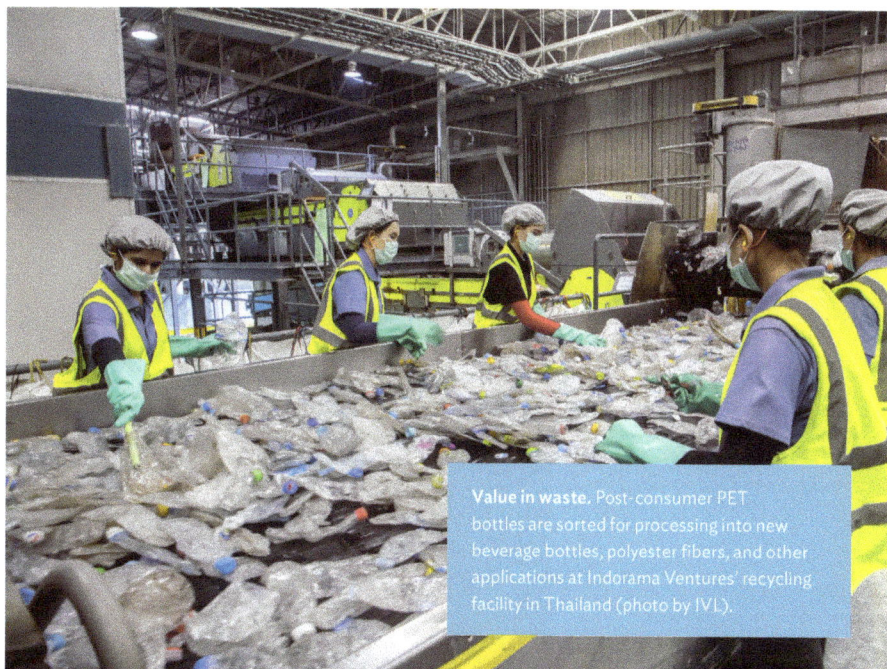

Value in waste. Post-consumer PET bottles are sorted for processing into new beverage bottles, polyester fibers, and other applications at Indorama Ventures' recycling facility in Thailand (photo by IVL).

Four Provincial Capitals, Three "Rs" of Recycling

Four provincial capitals will enact comprehensive solid waste management schemes for the first time, with a focus on the three "Rs": reduce, reuse, and recycle.

The Managing Solid Waste in Secondary Cities project intends to reduce environmental pollution, enhance community awareness, and increase the recycling of resources in the cities of Darkhan, Altai, Baruun-Urt, and Arvaikheer. This $2 million initiative is funded by the ADB-managed Japan Fund for Poverty Reduction, and will focus on the three "Rs" of reduction of waste at the source, reusing, and recycling. Pilot community practices for waste reuse will be implemented, and 15 waste recycling shops will be opened.

New controlled landfills and transfer stations will also be constructed, with clay lining, retaining walls, and diversion channels to minimize leakage and groundwater contamination. The new facilities, due to be completed by 2023, will be built with climate change adaptation in mind, with additional culverts and raised embankments in flood-prone areas.

An education campaign will be launched utilizing social media and other channels, and by the end of the project period in 2022, 24% of waste in targeted communities is expected to be recycled.

↑24%

OF WASTE IN TARGET COMMUNITIES IS RECYCLED

Inhabitants
Darkhan: 83,883
Altai: 17,617
Baruun-Urt: 21,550
Arvaikheer: 29,420

National GDP per capita
Darkhan: $4,295
Altai: $4,295
Baruun-Urt: $4,295
Arvaikheer: $4,295

Geographic area
Darkhan: 103 km²
Altai: 2,161 km²
Baruun-Urt: 52 km²
Arvaikheer: 173 km²

THE CHALLENGE

Waste generation increased seven fold from 2008 to 2017, but most cities still rely on open waste dump sites and only 1% of waste is currently reused.

CO-BENEFITS

Health

Reducing the open burning and disposal of waste in uncontrolled landfills will improve the reproductive and general health of residents.

Social

The project will improve employment in the informal trash collection and recycling sector, where individuals currently work in unsafe conditions at disposal sites.

Environmental

Environmental pollution from current waste management techniques, including groundwater contamination, will be reduced.

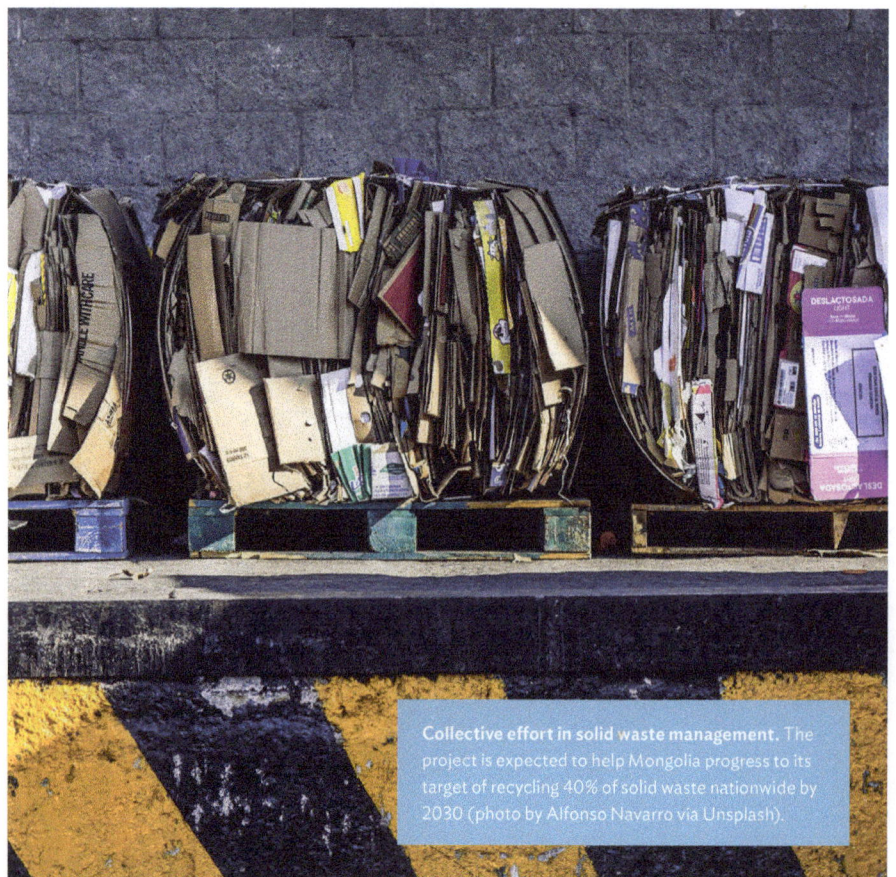

Collective effort in solid waste management. The project is expected to help Mongolia progress to its target of recycling 40% of solid waste nationwide by 2030 (photo by Alfonso Navarro via Unsplash).

A Fresh Approach to Food Waste in Ulaanbaatar

An ongoing community recycling project is exploring the untapped potential of food waste in Mongolia's capital.

The $3.2 million initiative, primarily funded by the ADB-managed Japan Fund for Poverty Reduction, works to turn food waste into compost through community participation, with pilot projects being implemented in households and schools within the city, and a large composting facility constructed under the project.

The participating households and schools will be provided with the necessary tools, including storage bins, composters, and protective equipment, to engage in composting on a small scale. The resulting compost will be used for green open spaces, flower beds, and vegetable gardens.

The new community composting facility will process at least 2,000 kilograms of food waste per day, and composters will be equipped with self-heating systems, which will ensure activities can be conducted even in the winter in the world's coldest capital. The composting facility is expected to become financially self-sufficient due to income generated from compost sales after only 1 year.

By 2024 it is expected that 700 tons of food waste will be composted each year, and GHG emissions will be reduced by 1,368 tons annually.

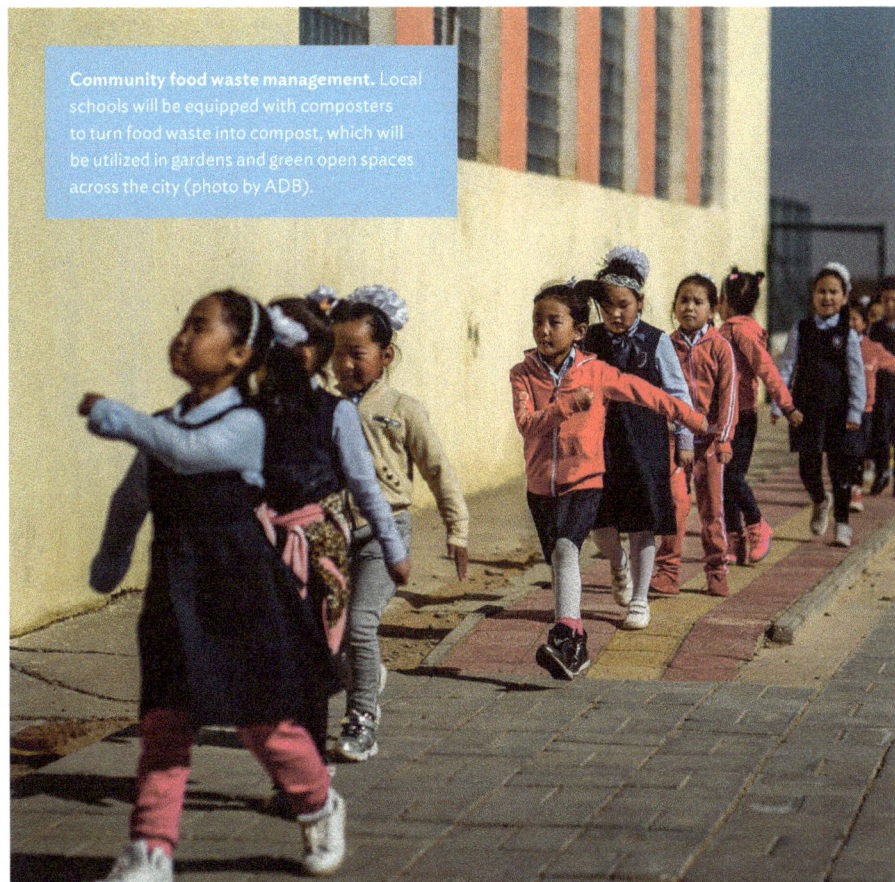

↓1.4K

TONS OF CO$_2$ EMISSIONS REDUCED ANNUALLY

Inhabitants
1.54 million

GDP per capita
$5,653

Geographic area
4,704 km^2

THE CHALLENGE

Ulaanbaatar generates more than 40% of Mongolia's solid waste, much of which is food waste that is illegally dumped or disposed of in landfills.

CO-BENEFITS

Economic

The project will provide opportunities for new businesses and jobs, particularly among vulnerable communities, with local schools, restaurants, and hotels all benefiting from project activities.

Environmental

The recycling of food waste through composting will improve environmental conditions, reduce health risks, and lower GHG emissions.

Community food waste management. Local schools will be equipped with composters to turn food waste into compost, which will be utilized in gardens and green open spaces across the city (photo by ADB).

Tackling Maldives' Trash Island

After 30 years of dumping solid waste, Maldives' capital city and surrounding islands are investing in a more sustainable waste management system to reduce emissions, improve resilience, and protect local fishing and tourism industries.

The Greater Malé region is on the way to establishing a more sustainable waste management system to replace a dumpsite on a nearby island that has been a health and environmental hazard for 30 years.

To improve waste management in the Greater Malé region and its neighboring outer islands, the Government of Maldives, through the Ministry of Environment, is implementing a two-phased project stretching to 2026. The project is financed through a combination of grants and loans from ADB, a loan from the Asian Infrastructure Investment Bank, government financing, and a $10 million grant from the ADB-administered Japan Fund for the Joint Crediting Mechanism, a special fund set up by the Government of Japan to incentivize low-carbon investments.

The first phase will invest $40 million in improved waste collection, transfer, and disposal, as well as awareness raising and behavior change around sustainable solid waste management. The second phase will invest a further $151 million in a climate-resilient waste treatment facility, including a waste-to-energy plant and improved landfill for treatment residues, as well as improving institutional capacity.

↓40K

TONS OF CO_2 EMISSIONS REDUCED ANNUALLY

MALDIVES

Inhabitants
533,900

GDP per capita
$10,791

Geographic area
300 km²

THE CHALLENGE

For the past 30 years, solid waste generated in the Greater Malé area has been dumped with no treatment and burned in the open on the industrial island of Thilafushi.

CO-BENEFITS

Social

The project includes improved waste management for poor outer island communities, including skills training and awareness and behavior change campaigns for sustainable solid waste management.

Health

Adopting more sustainable waste management practices will improve public and environmental health, especially for women and the poor.

Environmental

The project will decrease the amount of leachate and solid waste-like plastics that currently enter the sea from the trash island, improving ocean health.

Economic

The project will reduce ocean pollution and positively impact the tourism and fishery sectors, two cornerstones of Maldives' economy.

Maldives' waste management plans. The initiatives aim to completely rehabilitate Thilafushi, the current "trash island" where waste is being dumped (photo by Water Solutions Pvt. Ltd.).

Building a sustainable waste management facility. A graphical rendering of how the future waste management system could look like for Maldives on an area of rehabilitated land (photo by Water Solutions Pvt. Ltd.).

Climate Resilience

→ Cities are experiencing more disasters as a result of climate change, whether through severe flooding, heat waves, or droughts. Resilient infrastructure, adaptation plans, and nature-based solutions are all important tools demonstrated in this chapter that are helping cities reduce their vulnerability to natural hazards, while also leveraging co-benefits such as improved water and soil quality, carbon sequestration, and new recreational opportunities.

Nature-Based Solutions Enhance Resilience in Cities in Viet Nam

Several cities in Viet Nam have committed to a development approach that works with the surrounding environment to respond to challenges related to urban growth and climate change.

Viet Nam's largest urban center, Ho Chi Minh City, and the secondary cities of Vinh Yen and Hue are set to adopt water-sensitive urban design (WSUD) that integrates water management with urban development and the built environment.

WSUD's flexible nature is of particular use in this multi city context, as individual components can be optimized for each community. For example, city planners in Ho Chi Minh City are exploring options to mitigate flood risks by integrating floodplains and sloping topography in the large Go Vap Cultural Park. On a smaller scale, Hue City is experimenting with new types of edge treatments and WSUD tools to improve water quality along the Lap River and create a green community space.

Based on the project experience, these climate-resilient development approaches will be expanded to include other cities in Viet Nam in the future.

Funding is provided by the ADB-managed Urban Climate Change Resilience Trust Fund.

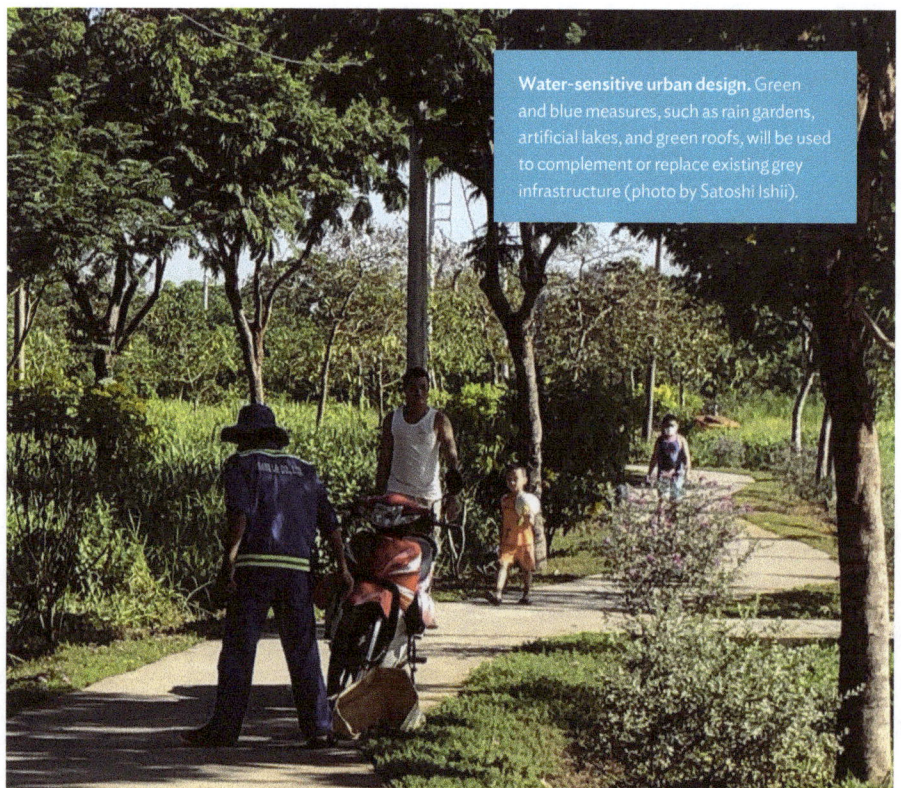

↑$4M
INVESTMENT IN CLIMATE RESILIENCE

Inhabitants
Ho Chi Minh City: 8.99 million
Vinh Yen: 152,801
Hue: 455,230

GDP per capita
Ho Chi Minh City: $6,862
Vinh Yen: no data
Hue: no data

Geographic area
Ho Chi Minh City: 2,061 km²
Vinh Yen: 51 km²
Hue: 71 km²

THE CHALLENGE

Growing cities in Viet Nam are often exposed to environmental degradation and are vulnerable to climate change-related hazards like flooding.

CO-BENEFITS

Environmental

WSUD approaches can improve water quality, prevent soil erosion and sedimentation, contribute to carbon sequestration, and increase biodiversity.

Social

By integrating natural components with the built city environment, WSUD will create a myriad of recreational use opportunities for communities.

Water-sensitive urban design. Green and blue measures, such as rain gardens, artificial lakes, and green roofs, will be used to complement or replace existing grey infrastructure (photo by Satoshi Ishii).

Flood Control in High-Risk Area of the Yangtze River Basin

↓40%

REDUCTION IN STORMWATER POLLUTANT LOADS

Inhabitants
1.19 million

GDP per capita
$11,815

Geographic area
3,178 km²

THE CHALLENGE

The capacity of existing water and waste infrastructure is insufficient to handle population growth and flooding in the watershed area.

CO-BENEFITS

Environmental

The project's green planning approach will reduce pollution from stormwater and waste, and will provide new green spaces for the public to enjoy.

Economic

A number of construction jobs will be created by the project, benefiting local communities.

Xinyu is adopting an integrated approach to flood control and waste management that seeks to safeguard communities against warming temperatures and increased precipitation.

The project aims to manage risks related to flooding and water pollution in the Kongmu River watershed, part of the larger Yangtze River Basin. The residents of Xinyu face flooding every year, which is expected to increase with climate change causing more precipitation in the region than the global average.[8]

A comprehensive strategy has been developed that focuses on increasing flood retention capacities, improving sanitation services for communities, and protecting Xinyu's water supply.

The city will increase its resilience to flooding through rainwater interception and other stormwater management systems, and constructed wetlands will help treat runoff and improve water quality, while also acting as natural flood barriers. Water regulation structures, such as levees and interconnected storage lakes, and a wastewater collection network that operates separately from stormwater collection will also be established.

The project has a budget of $300 million, provided in part by an ADB loan.

---

[8] Intergovernmental Panel on Climate Change. 2018. The People's Republic of China Third National Communication on Climate Change. December.

An integrated approach to flood control. The vulnerability of residents will be further addressed through the adoption of an early warning system and flood-resilient planning (photo by Ding Wei).

Coastal Towns in Viet Nam Prepare for a Warmer Future

Two cities along Viet Nam's vulnerable central coast are investing in climate-proof infrastructure and adaptation plans to improve their urban environments and climate resilience.

Dong Hoi and Hoi An, two important tourism centers that are located along some of Viet Nam's most disaster-prone coastal areas, are attempting to develop urban infrastructure that is sustainable and able to keep pace with rapid growth. The project will run until 2023, and efforts will center on climate-proofing existing and new developments through improvements to stormwater and flood management, erosion prevention, and salinity control.

Urban developments on the Bao Ninh Peninsula in Dong Hoi and along the Co Co River in Hoi An will include integrated flood management in the form of vegetated buffer zones to control water surges and sand dune restoration, an increase in the storage capacity of a reservoir, the establishment of a forecasting and warning system, and the creation of an evacuation route. The Lai Nghi Reservoir, an important water source, will also be protected against salinity intrusion, and zoning and vegetation will be introduced to protect coastal dunes from erosion.

The project has a budget of $144 million, $104 million of which is provided by ADB and the ADB-managed Urban Climate Change Resilience Trust Fund.

↓$2M

REDUCTION IN DAMAGE FROM COASTAL FLOODING PER YEAR

Inhabitants
Dong Hoi: 119,222
Hoi An: 152,160

GDP per capita
Dong Hoi: 3,871
Hoi An: no data

Geographic area
Dong Hoi: 156 km²
Hoi An: 60 km²

THE CHALLENGE

Dong Hoi and Hoi An have been threatened repeatedly by floods and typhoons. As rapidly growing midsize cities with tourism as their development drivers, both must ensure that urban infrastructure development considers climate resilience.

CO-BENEFITS

Economic

Coastal protection, improved roads, health services, water supply, and wastewater connections will result in savings for communities and increase the economic value of the area.

Social

Upgrades to urban infrastructure will allow for sustainable growth in tourism sectors, and real-time information will allow residents to quickly react to floods.

Climate proofing for development. Dong Hoi and Hoi An recognize the importance of sustainably building their tourism sectors and strengthening their capacity to respond to climate change (photo by ADB).

Urban Resilience on the Frontline of Climate Change

↑**99.9%**

OF POPULATION WITH
SECURE WATER SUPPLY

NUKU'ALOFA

Inhabitants
23,221

GDP per capita
no data

Geographic area
17 km²

THE CHALLENGE

Nuku'alofa's urban growth has not been matched by a similar increase in adequate public services, and the city faces dire climate change predictions.

CO-BENEFITS

Economic

The flood management aspects of the project will reduce flooding incidence and related repair costs for both households and businesses.

Health

Improved sanitation and hygiene will reduce exposure to waterborne diseases and improve water safety for residents.

Social

By implementing a disaster risk management program at the community level, organizers can target those most vulnerable to climate change impacts.

The ongoing urban resilience project aims to supply flood management, water and sanitation infrastructure, and a future resilience and climate strategy to the city's most vulnerable households.

Tonga's capital on the southern island of Tongatapu hopes to raise the standard of living for its 23,000 residents, while also enhancing the ability of the island to cope with natural hazards and effects of climate change, particularly flooding and sea level rise.

Eighty percent of Tonga's population and critical infrastructure are located in flat, low-lying coastal areas that are exposed to rising sea levels, with the frequency of flooding and prevalence of storm surges only expected to increase in the coming years. Some climate models suggest that without strong and fast climate action, most of current-day Nuku'alofa could be underwater by 2100.

Drainage systems will help improve the resilience of 40 hectares of land in seven flood-prone locations, while also contributing to public health efforts by reducing exposure to waterborne diseases. A reliable water supply will be secured for 99.9% of the population through upgrades to piped distribution networks, and the introduction of a long-term climate- and disaster-resilient urban development strategy and investment plan will increase the future resilience of the urban area.

The project is supported by an $18 million ADB grant.

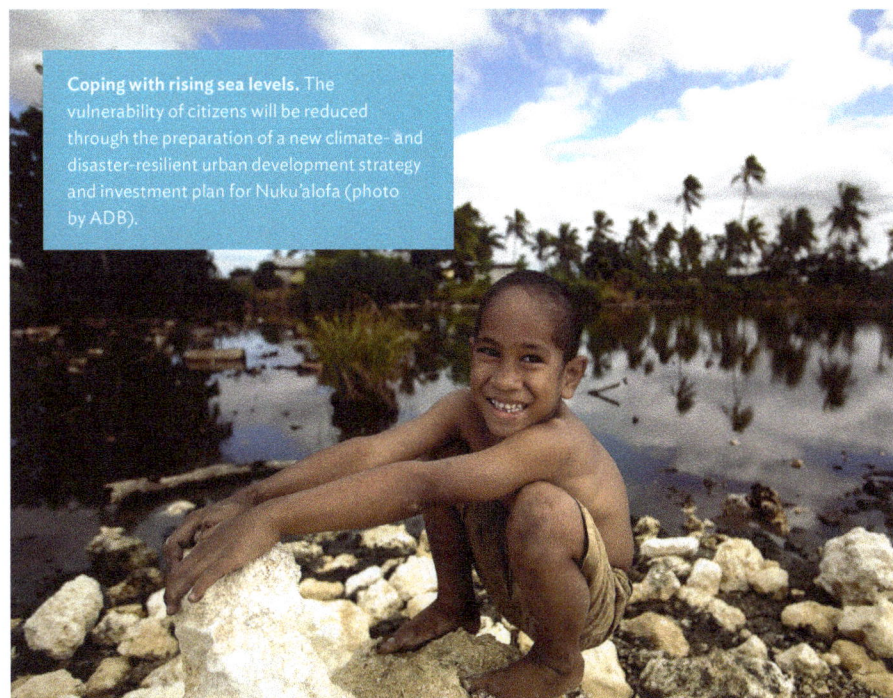

Coping with rising sea levels. The vulnerability of citizens will be reduced through the preparation of a new climate- and disaster-resilient urban development strategy and investment plan for Nuku'alofa (photo by ADB).

Nature-Based Solutions on the RISE in Makassar

↑1.6K

PEOPLE RECEIVE IMPROVED SANITATION AND WATER MANAGEMENT

Inhabitants
1.53 million

GDP per capita
$8,270

Geographic area
199 km²

THE CHALLENGE

Many of the 58,200 households who live in informal settlements lack access to centralized urban infrastructure and face threats from climate change.

CO-BENEFITS

Environmental

Reduced contamination and the restoration of natural processes will enhance the environment surrounding the settlements.

Social

Community-driven development approaches will empower beneficiaries of the RISE project and allow them to implement nature-based solutions that best fit their needs.

Health

Sanitation facilities and reliable water supplies will result in better community health, with fewer infections and water-related diseases and increased food production.

The pilot project will reduce the impact of climate change on marginalized communities in Makassar by providing localized, water-sensitive interventions that address issues related to sanitation and water management.

Running until 2022, the Revitalising Informal Settlements and their Environments (RISE) initiative will integrate nature-based solutions into water and sanitation cycles in the South Sulawesi Province of Indonesia. The program is conducted in partnership with Monash University, and is part of an ongoing environmental and human health research funded by a grant from the Wellcome Trust.

Six settlements that are home to around 1,600 people will receive location-specific solutions, including wetland restoration, bio-filtration gardens, stormwater harvesting, paved paths, and new sanitation structures to improve wastewater management and diversify water resources.

The introduction of water-sensitive solutions will enable targeted communities to recycle wastewater and rainwater, restore natural waterways, improve water quality, and reduce vulnerability to flooding.

Findings from the program will inform future climate adaptation efforts in urban water and sanitation, both within the region and globally.

Financing is provided through the ADB-managed Urban Climate Change Resilience Trust Fund and Southeast Asia Urban Services Facility.

Integrating water-sensitive solutions for improved livability. In a small settlement in the district of Batua, Makassar, Indonesia, the RISE program is using nature-based technologies such as this bio-filter to upgrade urban services (photo by RISE Program, Monash University).

Community-driven development. Co-designing infrastructure solutions together with the community, the RISE program constructed wetlands, bio-filtration gardens, rainwater harvesting, local sanitation systems, private toilets, and an access road (photo by RISE Program, Monash University).

Methodology

→ Sourcing Climate Actions from Cities in Asia and the Pacific

This publication has come to life as a result of a multistep process involving ADB, Sustainia, and on-the-ground sustainability experts from cities across Asia and the Pacific.

Finding the solutions:

Local consultants with expertise in sustainable urban development completed initial research on city projects and solutions. With guidance from ADB and Sustainia, the local experts then gathered information and data on the solutions from across Asia and the Pacific to aid the assessment and selection process.

Assessing and selecting the solutions:

Upon submission of the solutions, ADB's urban experts assessed each case based on the five selection criteria. With guidance from Sustainia, the solutions that best met the criteria were put forward to be featured in this publication.

HOW DID WE ASSESS THEM?

Each case has been assessed on the following five criteria:

1

CLIMATE ACTION

The expected or achieved CO_2 reduction and/or climate risk mitigation of the project

2

CO-BENEFITS

The extent to which the project has positive co-benefits (economic, environmental, health, and social) in addition to its climate change mitigation and CO_2 reductions

3

INNOVATION

The extent to which the project takes an entirely new or groundbreaking approach to address major environmental issues

4

GOVERNANCE

How well the project is incorporated into larger city plans, collaborates with other entities in the city, and engages citizens in the project's development and implementation

5

SHARING AND SCALING

The extent to which the project experience is shared with other cities and regions, and the future potential to scale the project within the city

Index

The check mark indicates which projects have received financial support from ADB.

www.ingramcontent.com/pod-product-compliance
Lightning Source LLC
Chambersburg PA
CBHW050043220326
41599CB00045B/7264